Palgrave Advances in Language and Linguistics

Series Editor
Jonathan Crichton
Communication, Int Studies and Languages
University of South Australia Communication,
Int Studies and Languages
Magill, Australia

This series was founded by Professor Christopher N. Candlin, and is now edited by Jonathan Crichton, Associate Professor in Applied Linguistics at the University of South Australia. It provides a home for research at the cutting edge of the discipline. It is part of an overall publishing program by Palgrave Macmillan aimed at producing collections of original, commissioned chapters under the invited editorship of distinguished scholars.

The books in the Series are not intended as an overall guide to the topic or to provide an exhaustive coverage of its various sub-fields. Rather, they are carefully planned to offer the informed readership a conspectus of perspectives on key themes, authored by major scholars whose work is at the boundaries of current research. What we plan the Series will do, then, is to focus on salience and influence, move fields forward, and help to chart future research development.

The Series is designed for postgraduate and research students, including advanced level undergraduates seeking to pursue research work in Linguistics, or careers engaged with language and communication study more generally, as well as for more experienced researchers and tutors seeking an awareness of what is current and in prospect in adjacent research fields to their own.

Editors of books in the Series have been particularly asked to put their own distinctive stamp on their collection, to give it a personal dimension, and to map the territory, as it were, seen through the eyes of their own research experience.

Larissa Semiramis Schedel • Cori Jakubiak
Editors

Voluntourism and Language Learning/Teaching

Critical Perspectives

palgrave
macmillan

Editors
Larissa Semiramis Schedel
University of Bonn
Bonn, Germany

Cori Jakubiak
Grinnell College
Grinnell, IA, USA

ISSN 2947-6623 ISSN 2947-6631 (electronic)
Palgrave Advances in Language and Linguistics
ISBN 978-3-031-40812-0 ISBN 978-3-031-40813-7 (eBook)
https://doi.org/10.1007/978-3-031-40813-7

This Palgrave Macmillan imprint is published by the registered company Springer Nature Switzerland AG.
The registered company address is: Gewerbestrasse 11, 6330 Cham, Switzerland

Paper in this product is recyclable.

To our research participants*—who were powerful collaborators.*
We are immensely grateful for your time and willingness to share
your experiences and points of view.

Praise for *Voluntourism and Language Learning/Teaching*

"This excellent volume exposes anew the entanglements of language, tourism, and neoliberal/ neocolonial capitalism. Each of the eight case studies reminds us how tourism discourse is to global inequality as "color blindness" is to racism: both are slights of hand conveniently serving the interests of the privileged. In this case, we witness travelers-by-choice exploiting their linguistic capital and reasserting their symbolic power, all under the earnest guise of philanthropy."

—Crispin Thurlow, *University of Bern, Switzerland.* Co-author, Tourism Discourse: Language and Global Mobility (Palgrave, 2010)

"Jakubiak and Schedel have put together an excellent collection of work that will be essential reading for scholars of voluntourism. In focusing on language as a motivation for travel, the authors, collectively and individually, have achieved that rare thing: a coherent edited collection that truly advances our understanding."

—Jim Butcher, *Canterbury Christ Church University, UK.* Co-author, Volunteer Tourism: The Lifestyle Politics of International Development (Routledge, 2015)

Contents

Notes on Contributors

Joshua D. Bernstein is a lecturer at the Language Institute, Thammasat University, Thailand. His research interests include the global English language teaching industry, alternative/sustainable tourism, and Southeast Asia (Laos and Thailand).

Carolyn Areum Choi (she/hers) is Assistant Professor of American Studies at Princeton University. She was previously a Korea Foundation Postdoctoral Fellow at the Center for Korean Studies at the University of California, Los Angeles. Her work examines how race, nation, migration, and political economy intersect, and her current book project looks at South Korean youth mobilities across intra-Asian and different Anglo-American settler contexts. She has published in *International Migration Review*, *Global Networks*, *positions: asia critique*, and *Sexualities*. When she is not writing for adults, Carolyn writes children's books on intersectional feminism and is co-author of *IntersectionAllies: We Make Room for All*.

Neriko Musha Doerr received a Ph.D. in Cultural Anthropology from Cornell University. Her research interests include the politics of difference, language and power, civic engagement, study abroad, and education in Japan, Aotearoa/New Zealand, and the United States. Her publications include *The Native Speaker Concept: Ethnographic Investigations of "Native Speaker Effects"* (Mouton de Gruyter), *The*

Romance of Crossing Borders: Studying and Volunteering Abroad (Berghahn; with Hannah Taïeb), *Transforming Study Abroad: A Handbook* (Berghahn), and *Fairies, Ghosts, and Santa Claus: Tinted Glasses, Fetishes, and the Politics of Seeing* (Berghahn), and peer-reviewed journal articles. She currently teaches at Ramapo College.

Cori Jakubiak is Director of the Center for Prairie Studies and Associate Professor of Education at Grinnell College in Grinnell, Iowa, USA. Her current research uses a critical sociolinguistics lens to examine two different kinds of language-based tourism: English-language voluntourism as a form of global citizenship education, and Italian language-based tourism as a mode of leisure, international mobility, and place-making. She teaches courses on the cultural politics of language teaching, differentiating instruction, and educational foundations.

In Chull Jang is an assistant professor in the Department of English Language Education, Seoul National University, South Korea. His research interests include critical sociolinguistic ethnography, student mobilities, and the English divide. His major publications have appeared in the *Journal of Sociolinguistics*, *Canadian Modern Language Review*, and the *International Journal of Bilingual Education and Bilingualism*.

Jessica McDaid is a PhD fellow at Universitat Autònoma de Barcelona and member of research project ENIFALPO, which studies Spanish family language policy choices regarding the acquisition of English. Specifically, she is interested in forms of adolescent mobility for English immersion purposes and the imaginaries and values guiding these decisions.

Kyoko Motobayashi is an associate professor in the Department of Language and Information Sciences within the Graduate School of Arts and Sciences at the University of Tokyo. Her main research areas are sociolinguistics, discourse analysis, and language policy studies. Her work has appeared in journals such as the *International Journal of Bilingual Education and Bilingualism*, *Multilingua*, and *Language and Sociocultural Theory*. Her most recent publication includes a chapter in *Discourses of Identity: Language Learning, Teaching, and Reclamation Perspectives in Japan* (edited by Mielick, Kubota, and Lawrence, 2022: Palgrave Macmillan).

Larissa Semiramis Schedel is a postdoctoral reseracher at the University of Bonn, Germany. She received a Ph.D. in Critical Sociolinguistics from the University of Fribourg, Switzerland. Her research investigates the various roles of language, the management of multilingualism, and the social consequences of linguistic commodification across different workplace settings in the tourism industry. Her current work focuses on voluntourism and English language travel in Malta.

Aviva Sinervo is Lecturer in Anthropology and Psychology at San Francisco State University, and in Human Development at the California State University, East Bay. Her research focuses on the moral and affective economies of child labor, volunteer tourism, international aid, and urban street selling, with a current focus on cosmopolitanism and generational-maturational shifts in wage-earning strategies in Peru.

Andrea Sunyol is a Margarita Salas postdoctoral fellow and a lecturer at UCL Institute of Education. Her research is centered around the study of multilingualism and social inequality, more specifically in the role of language in the educational strategies and practices of dominant social groups. She has conducted ethnographic research in elite education institutions in Catalonia and in pre-sessional EAP courses in Higher Education in the UK.

List of Figures

List of Tables

1

Introducing Language-Motivated Voluntourism

Cori Jakubiak and Larissa Semiramis Schedel

Setting the Stage

Over two decades ago, tourism scholar Stephen Wearing (2001) theorized the concept of *volunteer tourism* to describe an alternative way of travelling that combines international voluntary service with holidaying. Wearing's influential work set the stage for the emergence of scholarship on volunteer tourism, or what is referred to colloquially as "voluntourism" (see, e.g., Projects Abroad, n.d.). Recent scholarship has extended Wearing's original understanding of volunteer tourism as humanitarian

C. Jakubiak
Grinnell College, Grinnell, IA, USA
e-mail: jakubiak@grinnell.edu

L. S. Schedel (✉)
University of Bonn, Bonn, Germany
e-mail: lschedel@uni-bonn.de

L. S. Schedel, C. Jakubiak (eds.), *Voluntourism and Language Learning/Teaching*, Palgrave Advances in Language and Linguistics,
https://doi.org/10.1007/978-3-031-40813-7_1

or development-oriented North-South[1] leisure/service travel to include North-North/South-North/South-South mobilities (e.g., Chen, 2016; Sin, 2009) and activities that are not necessarily anchored in aid but rather in for-profit businesses (e.g., Brennan, 2014; Lyons & Wearing, 2012; Prince, 2017; Schedel, 2022). Thus, as we[2] develop further in this introduction, we view the category, *volunteer tourism*, as now including other international mobilities that merge travel and unpaid—or precarious—labor practices.

Language learning and language teaching have been identified as key motivational factors for individuals to engage in international mobilities that weave together volunteerism, work, and leisure. These mobilities include: humanitarian or development-oriented volunteer tourism (e.g., Jakubiak, 2020; Mostafanezhad & Hannam, 2014; Tiessen, 2018; Wearing, 2001); voluntourism as work (e.g., Schedel, 2022; Stainton, 2019); unpaid internships (e.g., Jang, 2017); working holidaymaking (e.g., Yoon, 2014); precarious employment in the context of gap year travel (e.g., Duncan, 2007); and seasonal labor within hospitality industries (e.g., Duncan et al., 2013). However, the specific role of *language* within these entangled work-leisure/tourism practices (Rice, 2010) merits more specific attention. This book contributes to such an effort, extending current voluntourism theorizing by critically examining the intersections among various forms of work/leisure travel and language learning/teaching. We thus view volunteer tourism and its cognates such as working holidaymaking, international internships, and gap year labor, among others, as discursive fields in which powerful ideas about

[1] We acknowledge that the terms, "Global South/North," (as well as "East/West") are complex, fuzzy, and contested. While we use these terms in this book, we remain unsettled by how categories such as "Global North/South" and others oversimplify complex spatialities (see Makoni & Pennycook, 2019, for a critical discussion of the varying geopolitical, metaphorical, and epistemological meanings of these concepts, as well as their heuristic utility within applied linguistics research).

[2] This edited collection has been produced collaboratively by Cori Jakubiak and Larissa S. Schedel. We equitably share the first editorship of the edited volume and the first authorship of the introduction as we have divided the labour and responsibilities as equally as possible. However, since academic publishing requires us to attribute a specific order of editor-/authorship (a hierarchical practice which has consequences in terms of the valuation of publications in scholarly careers), we have chosen to alternate the order of names for (1) the introduction and (2) the edited volume (see also Highet & Schedel, forthcoming).

language(s), their speakers, and pedagogical practices are propagated worldwide. In so doing, we seek to understand the (language and other) ideologies that undergird different forms of what we call *language-motivated voluntourism*: work/leisure travel that involves language learning or teaching in various emplaced contexts.

The contributions in this volume, however, do not measure the instructional impacts of language-motivated voluntourism or offer suggestions as to how language education within work-leisure/travel settings might be made more rigorous or efficient. Instead, this book examines the discourses that adhere to language learning and teaching—and the practices that follow from these discourses—within various forms of work/leisure travel. Taken as a whole, then, this collection attempts to shed new light on language practices at/as/about work (Boutet, 2021) and during leisure time in volunteer tourism spaces; the im/mobilities produced by language-motivated voluntourism; and the educational, social, and political-economic consequences for the individuals involved in language-motivated voluntourism practices. By adopting critical perspectives, moreover, our contributors highlight how linguistic ideologies as used within work/leisure practices relate to, (re)produce, or contest dominant formations, power structures, forms of subjectivity, and material and social inequalities.

By bringing together the interdisciplinary contributions of scholars from various fields such as anthropology, childhood studies, critical sociolinguistics, second language acquisition, sociology, and teacher education, this book uses a range of discourse analytic and ethnographic approaches to examine language-motivated voluntourism. Questions investigated by our contributors include but are not limited to:

- How do voluntourism promotional materials construct and marketize language learning/teaching abroad?
- How is language teacher preparation constructed and/or operationalized within voluntourism discourses and practices?
- What assumptions about language learning/teaching circulate among various actors in volunteer tourism formations?
- How is language learning/teaching experienced by voluntourists and those they aim to serve, the "voluntoured"?

- What kinds of language learning opportunities do different work-leisure/tourism practices afford?
- What linguistic practices can be observed by voluntourists as/at work during their service tenure?
- How is language learning/teaching shaped by governance structures at various scales in the context of work-leisure/tourism practices?
- How can linguistic commodification as an analytical lens contribute to deeper understandings of language-motivated voluntourism?
- What forms of im/mobility are enabled or constrained by language-motivated voluntourism?

By bringing together a collection of work by new and mid-career scholars whose research sites, taken together, include North America, South America, Europe, East Asia, and Australia, this edited collection aims to extend current understandings of the mechanisms of mobility and language flows (cf. Horner & Dailey-O'Cain, 2019) in a globalized, ever-changing world.

This introduction is organized from this point as follows. First, we provide a brief history of extant volunteer tourism literature and trace its present contours. When volunteer tourism is refracted through the lens of language learning/teaching, various forms of work/leisure travel emerge and prompt us to reconsider traditional understandings of volunteer tourism as always grounded in development or humanitarian aid (cf. Wearing, 2001). Next, we turn to the role of language within various voluntourism practices, drawing analytical support from work in critical sociolinguistics. Finally, after touching upon the role of criticality in this volume, we provide brief commentary on the different chapters and situate them within the book's themes.

Volunteer Tourism: A Changing Field

The last twenty years have been a rich and productive time within volunteer tourism scholarship. In his 2001 book, *Volunteer Tourism: Experiences that Make a Difference*, tourism researcher Stephen Wearing provided a baseline definition of the phenomenon that quickly became a classic in the field. In that text, Wearing characterized volunteer tourism as follows:

> The generic term "volunteer tourism" applies to those tourists who, for various reasons, volunteer in an organized way to undertake holidays that might involve aiding or alleviating the material poverty of some groups in society, the restoration of certain environments or research into aspects of society or environment. (2001, p. 1)

Wearing's framing of volunteer tourism tied the practice firmly to humanitarianism and/or development-oriented mobilities. However, unlike multi-year, aid-oriented international voluntary service programs such as the U.S. Peace Corps or Britain's Voluntary Service Overseas, volunteer tourism—in early research—was specifically linked to short-term, holiday-length travel (e.g., one week to six months; see Callanan & Thomas, 2005, for a discussion of "temporally deep" versus "temporally shallow" volunteer tourism). Another defining feature of volunteer tourism was that it rarely required volunteers to possess specialized skills or advanced professional credentials (Callanan & Thomas, 2005).

Researchers in the nascent field of volunteer tourism studies often sought to understand why increasing numbers of tourists worldwide were eschewing traditional vacations such as beach resorts and packaged holidays for volunteer opportunities in the Global South. Their research led to flurry of work in tourism studies that explored volunteer tourists' individual motivations (e.g., Lo & Lee, 2011); the distinctions among volunteering, tourism, and alternative travel (e.g., Lyons & Wearing, 2008); whether, or how, travel sponsored by non-profit organizations—such as volunteer tourism—distinctly challenged commodified mass tourism (e.g., Butcher, 2006; Wearing et al., 2005); and the links between volunteer tourism and ethical consumption (e.g., Butcher, 2003; Butcher & Smith, 2010). Much of this research focused specifically on the perspectives and experiences of volunteers, most of whom identified as White, female, under the age of 30, and from the Global North (TRAM, 2008), but not always (see Lo & Lee, 2011, for an examination of volunteer tourists from Hong Kong).

Many early studies of volunteer tourism celebrated the practice as decommodified, alternative travel that yielded benefits for guests and hosts alike. Research in this vein suggested that volunteer tourism had the potential to harness youths' civic engagement desires and strengthen civil society (e.g., McBride & Sherraden, 2007; McGehee & Santos, 2005). Related research offered that participating in volunteer tourism could lead to personal transformation (e.g., Mustonen, 2006; Wearing et al., 2008), and some studies stressed that volunteer tourism had rich experiential learning potential (e.g., Lyons & Wearing, 2008).

Critical studies of volunteer tourism emerged swiftly as the field of study expanded. Kate Simpson's (2004, 2005) work was pivotal in this regard; she drew attention to the ways in which the expanding volunteer tourism industry—comprised for-profit operators in addition to non-governmental organizations (NGOs)—used Orientalist tropes in their marketing materials that created a discursive "geography of need" (see also Keese, 2011, on the geographical imaginaries of volunteer tourism). Simpson also disparaged the commercial gap year market for encouraging young, inexperienced volunteers to take on professional identities that they did not possess, such as *medical worker* or *teacher*, and "to experiment with [these] identit[ies] *on* a group of people" (2005, p. 466, emphasis in original). In that sense, early critical research on volunteer tourism explored a nontrivial question: Who are the primary beneficiaries of volunteer tourism: volunteer participants, host communities, or another entity entirely?

More recent volunteer tourism scholarship has more fully explored this question of *cui bono*. Drawing upon Foucauldian theories of governmentality as well as Anthony Giddens' (1994) work on life politics, numerous scholars have argued that volunteer tourism operates primarily as a form of geopolitical soft power (Nye Jr., 2004) and biopolitical regulation in an era of state retrenchment (cf. Henry & Mostafanezhad, 2019). Other scholars have offered that volunteer tourism reflects and upholds neoliberal exigencies and prerogatives. As gaining purchase on a middle-class future becomes increasingly elusive, more youth are seeking novel forms of cultural capital, such as a volunteer tourism experience, to accrue personal distinction (e.g., Vrasti, 2013). At the same time—and also due to post-Fordist economic restructuring—fewer state-sponsored social safety nets exist worldwide, which has naturalized the idea that civil society rather than welfare policies should address social problems (cf. Mostafanezhad & Hannam, 2014). The conflation of these two political-economic realities—(1) contemporary youths' needs to accrue distinctive forms of cultural capital in a time of middle-class enclosure; and (2) a diminished public sector—have codified neoliberal expectations that social problems are best addressed by civil society, ethical consumption, and individuals' efforts rather than policy amendments (see Butcher & Smith, 2015, for a discussion of how individual "life politics" has displaced the Fordist state). Among many critical volunteer tourism scholars, then, the ultimate beneficiaries of volunteer tourism may be the neoliberal structures and formations that generate and sustain both North-South and intranational inequality. As political theorist Wanda Vrasti (2013) observes about volunteer tourism's contradictions, "young adults joining these trips are ultimately also victims of larger socio-economic transformations (…) [they are] desperately trying to live up to the requirements of neoliberal subjectivity" (p. 132).

A groundswell of public critique against volunteer tourism has also emerged in the last few years. Well-aware of these concerns, many current volunteers and volunteer tourism sponsoring organizations now attempt to distance themselves from charges that these programs are neocolonial, dismissive of local communities' wants and needs, and ineffective in terms of outcomes (see Schwarz, 2018). On-line groups such as *Barbie Savior* and the Tumblr *Humanitarians of Tinder* have used parody to mock volunteer tourists who exploit vulnerable peoples' images for social media purposes (Schwarz & Richey, 2019; Wearing et al., 2018), while journalist Teju Cole's (2016) framing of international volunteering as part of a broader "White Savior Industrial Complex" has been taken up by many public writers. Despite some scholars' assertions that volunteer tourism can offer equitable opportunities for cross-cultural or linguistic exchange (e.g., Everingham, 2018), others have gone so far as to call for the end of humanitarian-oriented volunteer tourism entirely, given its inequitable power relations and overwhelming Whiteness (see Henry, 2018a, b).

From Segmented Volunteer Tourism to Language-Motivated Voluntourism

In the face of burgeoning scholarship on volunteer tourism—particularly within tourism studies—researchers have called for more segmentation of the phenomenon. That is, the concept of volunteer tourism as an alternative travel practice that combines humanitarian or development-oriented volunteering with holidaying (cf. Wearing, 2001) may no longer wield the heuristic power that it once did. What volunteer tourists actually *do* on their service projects varies widely: volunteers engage in ecological conservation work; assist with scientific research; do light construction projects; distribute supplies in refugee resettlement settings; support medical clinics; assist animals in rehabilitation centers; and teach English, among many other activities (TRAM, 2008). Grouping all these pursuits together as simply "volunteer tourism" may be of diminished analytical use, particularly as different volunteer projects have distinctive geopolitical footprints (Henry & Mostafanezhad, 2019). Tourism scholar Hayley Stainton (2016) speaks to the need for volunteer tourism segmentation in this way:

> [It] is time that the different segments [of volunteer tourism]—teaching, marine, conservation, research and so forth, are segregated in order to facilitate a more accurate and thorough comprehension …*[V]olunteer tourism should no longer be viewed as a single tourism form, rather it should be viewed as a macro umbrella term encompassing a number of micro-niches.* (p. 258, emphasis added)

In other words, grouping together family voluntourism (e.g., Germann Molz, 2021); volunteer tourism in animal sanctuaries (e.g., Parreñas, 2016); infrastructure-related volunteer tourism, such as building a well (Park, 2018); refugee resettlement volunteer tourism (e.g., Knott, 2018); and English-language voluntourism (e.g., Jakubiak, 2020) renders analysis of a singular "volunteer tourism" quite a challenge. Moreover, whether or to what extent these various projects constitute humanitarianism or development assistance—a longstanding assumption made about volunteer tourism—is another ongoing debate in related literature (see Butcher & Smith, 2010; Jakubiak, 2016a; and Palacios, 2010, among others).

Jacob Henry and Mary Mostafanezhad's (2019) entry in the *Handbook of Tourism and Globalization*, "The Geopolitics of Volunteer Tourism," is instructive on this point. They consider three qualitatively different kinds of volunteer tourism—teaching, environmental conservation work, and infrastructural projects—from three distinct geopolitical perspectives: feminist/everyday, environmental, and imperial. Teaching in the context of volunteer tourism, they note, exacerbates gendered ideas of labor. In the context of volunteer tourism, underprepared volunteer teachers—usually women—insert themselves into Global South classrooms and often resort to feminized care work in lieu of rigorous content area instruction, which they are ill-qualified to enact (see Jakubiak, 2016b). In a different vein, environmental conservation work in volunteer tourism is ideologically aligned with Western constructions of nature that prioritize enclosure and take a binary view of human and non-human animal communities (e.g., Parreñas, 2016). Finally, Henry and Mostafanezhad observe that infrastructural projects within volunteer tourism—such as building roads or wells—call to mind James C. Scott's (1998) arguments about "seeing like a state": they aim to make people (and their practices) legible to actors whose power resides outside of local communities. Thus, volunteer tourists engaged in infrastructural projects may prioritize objectives such as "modernizing" water systems with little regard to how these projects not only alter communities but also make them more amenable to regulation and surveillance from afar. Apparent here is how all three of these different volunteer tourism projects—teaching, environmental conservation, and infrastructural efforts—require distinct theoretical tools for making sense of their purposes and (intended or unintended) outcomes.

It is at this juncture that we put forward the concept of *language-motivated voluntourism*. Informed and inspired by scholarship in mobility studies (e.g., Collins & Shubin, 2015; Sheller, 2014), working holidaymaking (cf. Yoon, 2014), work/leisure programs such as Workaway (e.g., Brennan, 2018; Dlaske, 2016; Schedel, 2022), and language travels (e.g., Chun & Han, 2015), we define language-motivated voluntourism as travel spurred by language learning/teaching that involves precarious or unpaid labor as well as leisure. Language-motivated voluntourism thus includes English language teaching in the context of humanitarian or development-oriented volunteer tourism (Bernstein, this volume; Jakubiak, this volume; Sinervo, this volume); voluntary heritage language teaching in overseas contexts (Motobayashi, this volume); precarious language assistantship work among gap year youth (McDaid & Sunyol, this volume); and unpaid or precarious labor conducted in the pursuit of language learning (Choi, this volume; Jang, this volume; Schedel, 2022, this volume). All these forms of work/leisure travel involve voluntary or temporary labor practices, leisure, and language learning or teaching (and sometimes many of these—see Sinervo, this volume).

To provide more context on language-motivated voluntourism, we now turn to a discussion of language—particularly as it intersects with tourism studies.

Language in Tourism

Scholars from various disciplines have traced the broad role of *language* in tourism fields. In an overview article, Gavin Jack and Alison Phipps (2012) distinguish three approaches to the study of language in tourism contexts. The first is the *tourism impact approach*, which draws on a social psychological perspective to investigate the effects of tourism on attitudes toward local languages. The second is the *language of tourism approach*, which views tourism as a discursive phenomenon and explores how tourism destinations and experiences are discursively and semiotically produced in touristic brochures, the narratives of tour guides, travel blogs, etc. (e.g., Jaworski & Pritchard, 2005; Thurlow & Jaworski, 2011). This second approach focuses on the discursive characteristics of touristic texts or on what Graham Dann (1996) called "the language of tourism." The third approach is termed *intercultural communication in tourism*; this approach investigates dimensions of linguistic accommodation in the interactions between tourists and hosts as well as the language ideologies that underly users' linguistic practices. More recent work by Bal K. Sharma and Shuang Gao (2022) draws from this third approach, deploying a more critical conceptualization of intercultural communication that follows Ingrid Piller (2017), among others. The individual contributions in Sharma and Gao's (2022) edited volume highlight the dynamic construction, staging, and performance of linguistic and cultural differences in touristic contexts.

We believe that these three approaches to the study of language in tourism contexts (following Jack & Phipps, 2012) should be extended by a fourth one—the *linguistic commodification approach*—which investigates how languages are treated as economic resources in the tourism industry with numerous consequences for the involved speakers/users (see Duchêne, 2009), as tourism workers are often also "language workers" (Duchêne, 2019) and/or "commodified persona" (Bunten, 2008). The tourism industry is a service sector *par excellence* where language(s) get commodified in multiple ways: for instance, as working tools for communication with different markets of tourists (e.g., Duchêne, 2009), with English as a lingua franca/"international language" as a highly valued tool (e.g., Wilson, 2018). Scholars have examined how the tourism industry constructs the linguistic needs of different markets of tourists and accordingly trains future tourism workers in these languages (e.g., Duchêne & Piller, 2011) and employs workers with specific linguistic repertoires (e.g., Schedel, 2018a). However, scholars have also demonstrated that this linguistic accommodation/service is not always valorized as imagined (e.g., Muth, 2018; Schedel, 2023). As market constellations are ever-changing, languages can have fluctuating value; consequently, investments in particular languages are speculative (Duchêne & Daveluy, 2015). Language gets further commodified in tourism as an index of locality, authenticity, and exoticism (e.g., Heller, Jaworski, & Thurlow, 2014; Pietikäinen & Kelly-Holmes, 2011; Salazar, 2005) or as part of the touristic experience itself (e.g., Schedel, 2018b).

Linguistic commodification is highly salient in the case of language-based tourism, where language appears to be peoples' core motivation for travelling. Tourist language learners consume language as a product in formal language classes or immersion experiences, which also "sell" and market particular languages (Iglesias, 2016; Yarymowich, 2004). While the desire for English and other so-called global languages constitutes a billion-dollar industry and makes the language holiday an important segment of the tourism industry (Iglesias, 2016), less-commonly taught languages also get commodified for diaspora tourists and other holidaymakers (cf. Drozdzewski, 2011; O'Rourke & DePalma, 2017). Local languages hence function as attracting, but also potentially off-putting, factors when it comes to tourists' choices of destination (see Cohen & Cooper, 1986; Laesser et al., 2014). In geographical terms, however, language-based tourism does not always involve the crossing of national borders. Various types of intranational language-based tourism also exist. Tourists sojourn to particular regions within a nation-state where a minoritized language is used (e.g., Marty Crettenand, 2016), or they participate in language-focused summer camps (e.g., Petit Cahill, 2020) or language villages—specific enclaves where target languages are spoken for educational purposes (e.g., Lee, 2011). Shuang Gao's (2019) work in this area is particularly compelling; she explores how language becomes entangled in the dynamics of the social transformation of a touristic village in China—a village that figures as a destination for Chinese tourists who aspire to learn English through immersion but lack the economic resources to go abroad.

Still, language learning in the context of tourism has received surprisingly little academic attention. Notable exceptions are Crispin Thurlow and Adam Jaworski (2010), who demonstrate how guidebook glossaries invite tourists to learn and use phrases in local languages, thereby commodifying the local "linguaculture" (see also the 2014 special issue edited by Jaworski et al.). Alison Phipps (2006), taking a different tack, uses a phenomenological, auto-ethnographic approach to reflect upon her own language learning experiences during a language holiday with language courses and a family home stay in Lisbon. Also, scholars have examined Grand Tour sojourners' language learning activities (e.g., Cohen, 1996; Tosi, 2020). Moreover, while contemporary study abroad programs for traditionally-aged college or university students can be described (in some ways) as a form of language-based tourism (e.g., Iglesias, 2022), research at the intersection of language and study abroad has tended to use an identity lens rather than a tourism one (cf. Goldoni, 2013; Kinginger, 2004). A key exception is Neriko Musha Doerr and Hannah Davis Taïeb's (2020) edited collection, which investigates the romantic (touristic) imaginaries of people, landscapes, and languages held by travelers studying or volunteering abroad.

Within work/leisure mobility studies, language learning has been discussed as a motivational factor for tourists to engage in international volunteer work (e.g., Brennan, 2014, p. 262; Jang, 2017, 136–139; Takahashi, 2013, p. 68; Tiessen, 2018, p. 35). However, only Phiona Stanley (2017) provides a detailed auto-ethnographic analysis of learning Spanish while traveling and volunteering in South America. Relatedly, Larissa Schedel's (2022) study of how learners of English get exploited as a cheap labor in Malta's voluntourism industry is one of very few works of a similar theme.

Like language learning, language *teaching* also constitutes a central but heretofore narrowly explored activity within the field of touristic mobilities. There are, for instance, numerous professional teacher mobilities that involve touristic activities (e.g., Codó, 2018; Rey et al., 2020), and many travelers who have not received preparation in language pedagogy are able to teach abroad due to "native speaker" ideologies (see Doerr, 2009; Phillipson, 1992). This latter phenomenon occurs routinely in the context of global English language teaching; studies of native speakers teaching English in tourism contexts have been theorized in studies of "English-language voluntourism" (e.g., Jakubiak, 2016a, b, 2020; Thomas-Maude et al., 2021), "TEFL (teaching English as a foreign language) tourism" (Stainton, 2016, 2018, 2019), "TEFL voluntourism" (Bernstein & Woosnam, 2019), voluntary English teaching (Bunce, 2016), and "backpacker teachers" (Stanley, 2013).

Following this line of research, scholars have questioned the discursive and material effects of short-term, English language teaching/travel programs, which tend to employ underprepared, uncredentialled teachers and often lack formalized curricula (Bernstein, this volume; Bunce, 2016; Henry, 2020; Jakubiak, this volume; Sinervo, this volume). Robert Phillipson's (1992) work on "linguistic imperialism" has been vital in helping scholars to unmask the romanticization of White/Western savior narratives in short-term, volunteer global English teaching (e.g., Bunce, 2016; Jenks & Lee, 2020), as well as the discourses that adhere to it (see Jakubiak, 2012, e.g., on hyperglobalism in English-language voluntourism). Taking a hopeful perspective, Cori Jakubiak has pointed out that some volunteers develop *critical language awareness* (Alim, 2005) through their short-term, volunteer English teaching experiences abroad (2021, this volume). Still, Pauline Bunce (2016) warns that short-term, voluntary overseas English language teaching can enable child exploitation, as this largely unregulated practice gives potential sexual predators easy access to vulnerable children in schools and orphanages.

While studies of language teaching in voluntourism contexts have focused almost exclusively on English, languages other than English are also taught within language-motivated voluntourism contexts. For instance, Kyoko Motobayashi (2015, this volume) traces the contours of an international state volunteer program that sends Japanese volunteers to teach Japanese to *Nikkei* diaspora communities in South America. Also, Jessica McDaid and Andrea Sunyol (this volume) demonstrate how European gap year students' abilities to offer multilingual language support to international students in Northern Ireland boarding schools legitimize these gap year students' precarious labor. Thus, a small but growing body of work has begun examining the intersections among languages other than English, language teaching, and work/leisure travel.

In sum, language learning, language teaching, and language itself all swirl mightily within tourism mobilities. And while many observers take a celebratory view of this swirling—praising the ways in which global language flows ease and even produce new touristic mobilities (cf. Castillo Arredondo et al., 2018)—others take a distinctly more cautionary stance. This book's contributors align with this latter view; so, it is to a discussion of criticality that we now turn.

Critical Perspectives

The project of social critique aims to reveal harmful social practices and prompt change (e.g., Haslanger, 2021). Yet, the role and involvement of academic scholars in being allies to oppressed groups and working to achieve social equity/justice—if possible—is debated (e.g., Heller, 2001; Urla, 2021). While a Foucauldian- (e.g., 1991) inspired critique will be content with questioning taken-for-granted assumptions and unveiling the workings of governance—not necessarily providing solutions—there are other approaches to scholarly work that formulate recommendations or intentionally engage in activism. The latter not only seek to unveil the unequal distribution of and access to material and symbolic resources (following Bourdieu) but also attempt to achieve a redistribution of resources (for a more in-depth discussion of redistribution in social justice projects, see Fraser & Honneth, 2003).

As pointed out previously in this introduction, various scholars have started to critically revisit voluntourism practices by mobilizing different analytical lenses (feminist, postcolonial, anti-racist, new materialist, etc.). In this volume, we extend these critical perspectives to language-motivated voluntourism. To do this work, we have chosen to *center* and, at the same time, *decenter* language (Tupas, 2020) in our study of voluntourism. We next explain what we mean by centering and decentering language—which, at first glance, might seem to be quite a contradictory endeavor. We also characterize our project of critique.

First, we have chosen to center language in our research on voluntourism to direct the attention of tourism scholars, applied linguists, and educational practitioners to the understudied phenomenon that we call "language-motivated voluntourism." When looking at the increase and diversification (or segmentation, as Stainton, 2016, puts it) of the various mobilities that encompass volunteering abroad and touristic activities, we argue (along with other scholars, see previous section on "Voluntourism") that the original definition of voluntourism provided by Wearing (2001) falls too short/is outdated. However, from a critical perspective, we are not only concerned about the accuracy and timeliness of the definition of voluntourism but also its exclusionary effects. In a reflection on the practice of social critique written with co-authors, Jacqueline Urla rightly states that "definitions are inherently exercises of power" (Highet et al., forthcoming). When producing knowledge on voluntourism, then, we are compelled to ask: Who gets to decide what is understood as voluntourism (or not); which social practices and actors fit the definition and attract researchers' attention (or not); and whose voices get amplified (or not)?[3]

[3] See also Nermin Ismail's (2015) work on this point.

Importantly, by including stories of people whose im/mobilities have often been underexamined in volunteer tourism studies (e.g., children; people who are not read as White/Western; illegalized migrants; learners/teachers of languages other than English; etc.), this book intends to pluralize/diversify voluntourism narratives and prevent what novelist Chimamanda Ngozi Adichie (2009) called in her TED talk "the danger of a single story." This decentering, polyphonic approach—that is, our privileging of the voices of marginalized people—volunteers, volunteered, as well as the voluntoured[4]—whose perspectives have been less considered in the literature—is also an attempt to decolonize practices of language learning and teaching (cf. Browning et al., 2022) in the context of voluntourism. To allow for deeper insights into contemporary practices of (language-motivated) voluntourism, we have hence gathered several contributions that adopt ethnographic approaches and use "thick description" (Geertz, 1973) of the trajectories and experiences of research participants. Also, in order to interrogate the dominant discourses circulating in the context of language-motivated voluntourism, other contributions in this book employ discourse analytical approaches to reveal voluntourism's underlying (linguistic and other) ideologies.

Centering language further means to consider "the language part" (Boutet, 2001; see also Moïse & Wilson, 2023) of voluntourism, which has long been overlooked in related research. This oversight is reflected not only in terms of the only relatively recent interest among scholars in the activities of language learning and teaching within voluntourism but also regarding the role that language plays more generally in touristic mobilities. There is so much going on language-wise; yet, language has never figured prominently in work/leisure mobility studies *writ large*. This means that scholars have missed opportunities to use language as a lens for examining (or listening to?) the wider social and

[4] By "volunteers," we mean travelers who voluntarily engage in volunteer work, whereas we call "volunteered" those who wish to work, but who are not granted the possibility of doing paid work and end up accepting working for free/on a voluntary basis (see, e.g., Jang, this volume; McDaid & Sunyol, this volume). By "voluntoured," we refer to the host communities that get "toured" by visitors and, in the case of language-motivated voluntourism, receive language instruction from visiting volunteers or are used as source of "authentic" language and as learning partners by visiting volunteers who wish to learn a host community language (see Everingham, 2018, as well as Sinervo, this volume).

politico-economic processes that play out in the context of voluntourism practices. This volume, then, attempts to amend this gap. We take our inspiration from the critical sociolinguistic perspective (cf. Boutet & Heller, 2007; Heller, 2002, 2010), which highlights how language produces and shapes structures of power, relations of inequality, moralizing ideologies, and hegemonic ideas about racial identity (see also Alim et al.'s, 2016 compilation on raciolinguistics, which does similar work).

(Volun)tourism is indeed a setting that is characterized by unequal encounters. As Jack and Phipps (2012) note, tourism interactions "take place in contexts of power and status differentials that reproduce symbolic (and economic) inequalities between interlocutors" (p. 537). Elisa Burrai and her colleagues (2017) add that tourism should be understood "as a series of relational, processual, unequal and mobile practices" (n.p.), and Mary Mostafanezhad and Kevin Hannam (2014) assert that tourism encounters are also "moral encounters." Looking at these unequal encounters through the lens of language means to investigate how semiotic resources get conceived as named languages that are (together with their speakers) (de)valued and hierarchized under specific historical and political-economic conditions. However, as Ruanni Tupas (2020) reminds us, "[W]e [scholars] must not assume that language is central to speakers' lives as is usually the case when we study the role of language in society" (p. 228). He therefore suggests decentering language and looking "beyond language itself to make sense of how language is used" (p. 228), with what meaning, effects, and consequences for whom. This shift of the focus from the study of language (structures and use) to the study of its embeddedness in material (also in terms of spatial/geographical, cf. Park & Gao, 2015) and symbolic conditions in society joins what Jan Blommaert (2001) argued in his article with the title, "Context is/as Critique": namely, not to overlook the contexts in which language practices take place, get shaped, and are constrained.

In our attempt to decenter language in the study of language-motivated voluntourism, we also follow Monica Heller, Adam Jaworski, and Crispin Thurlow (2014, p. 427), who view tourism as a site of analysis for the exploration of wider social, economic, and political processes of late modernity, in the sense of what Jan Blommaert (2010) called a "sociolinguistics of globalization." Drawing on Arjun Appadurai's (1996) understanding of globalization as an ever-increasing circulation of things, money, ideas, and people, we must acknowledge that language also travels (cf. Chun & Han, 2015; Park, 2009; Pennycook, 2020; Wilson, 2016). People travel abroad to teach their own language(s) or go abroad to learn new ones; they learn languages "accidentally" while being on the move; and people bring languages back home in the form of "truncated" multilingual repertoires (Blommaert, 2009). The linguistic hierarchies that make people desire a language and move (physically or not) to acquire it, or the languages that people experience in various linguistic communities of practice, are "embedded in larger social forms of inequalities which are colonially-, class-, and ethnolinguistically shaped" (Tupas, 2020, p. 233). In this volume, we explore the relations among language learning/teaching and political economy in the context of voluntourism, asking what makes people aspire to learn or to teach a language abroad, under what circumstances, for what price, and with what consequences.

Outline of the Book

The first four content chapters of this book are organized under the heading, "Language-Motivated Voluntourism in Contexts of Leisure and Holiday Travel." The opening content chapter, contributed by Cori Jakubiak, examines English-language voluntourism by drawing upon data collected through a discourse analysis of NGO sponsors' promotional websites and interviews with in-service and former volunteers. Her analysis highlights four different discourses that actors use to justify volunteers' interventions in Global South classrooms. One of these discourses, *native English immersion as pedagogy*, reflects immersion as linguistic ideology as well as *native speaker saviorism* (Jenks & Lee, 2020).

Joshua Bernstein's contribution provides an analysis of on-line TEFL certification programs sold through Groupon's global e-commerce marketplace. These online programs are marketed to aspiring or current volunteer tourists, who often lack any relevant teaching credentials besides their (native) English language speaker status. Using TESOL International Association's standards as a lens to analyze these programs, Bernstein reveals how online credentialing may further commodify language education and perpetuate unequal power dynamics between volunteer teachers and their students.

Aviva Sinervo's chapter takes us to Peru, where she examines the complexities of humanitarian-oriented volunteer tourism in historical Cusco. There, bidirectional language exchange between visiting volunteers and local child vendors is common. Volunteers enroll in Spanish language schools, which also provide them with opportunities to teach English to children in afterschool and social service programs. Through vivid ethnographic vignettes, Sinervo reveals how children often prefer more relational, one-on-one English language practice to the more formal, pen-and-pencil-based, instruction that visiting volunteers provide. Her findings shed light on the tensions that emerge when children's agency rubs up against volunteer teachers' understandings of what "meaningful service" through language education should look like.

From Peru, we move to Japan: Kyoko Motobayashi's chapter examines the off-duty expectations of three international volunteer Japanese language teachers (JLTs) who were soon to depart for South America, where each volunteer was under contract to teach Japanese as a heritage language in a different *Nikkei,* or Japanese descendant, community through a government-sponsored program. Volunteers' destination choices were largely based on their expectations for language study outside of work. In that sense, these volunteers, or "middling transnationals" (Clarke, 2005), situated their volunteer language teaching within their own imagined mobile futures—futures in which knowing particular languages had more or less perceived exchange value.

In the second content section of the book, "Language-Motivated Voluntourism as Precarious Labor," the chapters consider two broad themes: (1) how peoples' desires for particular languages and global experience make them vulnerable to predatory employers and/or profit-making enterprises; and (2) how peoples' abilities to marshal linguistic resources can also expose them to exploitation. Taking up these themes, Larissa Schedel's chapter focuses on the example of a young man from Northwest Africa who aspires to learn English to fulfill his dream of becoming an entrepreneur in Europe. Having entered Europe as a language student, he uses voluntourism as a pure survival strategy after his visa expires. Tracing his five-year odyssey through Europe, Schedel uses ethnographic and interview data to explore how this young man's language learning trajectory and voluntourism practices are shaped by neoliberal governance. While his dream of making a living in Europe is undergirded by the desire for upward social mobility and emancipation from cultural constraints in his country of origin, he ends up submitting himself to extremely precarious living and working conditions.

In Chull Jang's chapter focuses on Canada, where Korean youth engage in language tourism in private English schools that also oversee supplemental internship opportunities in greater Toronto. Using the concept of migration infrastructure (Xiang & Lindquist, 2014), Jang's research investigates the roles that intermediaries play in institutionalizing internship experiences for Korean students who are studying English. Findings suggest that internships are becoming increasingly professionalized and commercialized. While this increased professionalism led many Korean English language learners in Jang's study to complete internships, others were dubious about the value of expending their labor power for uncertain (linguistic) returns.

Carolyn Choi's contribution to the volume situates the everyday experiences of South Korean working holidaymakers in Australia in the larger contextual narrative of South Korea's global ascendance. While South Korea's repositioning as a global power opened doors to liberalize travel and other forms of global citizenship, Choi's analysis shows how working holidaymakers' experiences on the ground often tell a different story that expresses the limits of South Korea's national power. While young South Koreans travel to Australia in hopes of learning English and gaining cultural exposure, instead they find themselves conscripted into a migrant labor system that precariously positions them as unfamiliar outsiders. Drawing upon ethnographic data, Choi argues that these migrant workers bargain for paradoxical inclusion as they accept their racialized positioning in Australian society as a way to carve out limited, albeit cosmopolitan, experiences—including access to English. This paradox illuminates how race limits the social standing of South Koreans despite their country's economic ascendance.

Jessica McDaid and Andrea Sunyol's chapter then takes us to Northern Ireland, where state-funded (public) boarding schools now serve as an enticing destination within the language tourism boom streaming from

Europe (Codó & McDaid, 2019) and Asia (Song, 2011). Using ethnographic data, McDaid and Sunyol show it is not just the student body that is changing within these historic institutions but also the landscape of their workers. Allured by romantic promises of immersion (Doerr & Taïeb, 2020) and the potential for self-capitalization, a unique category of language-and-tourism workers has emerged in Northern Ireland's boarding schools. "Gaps," as they are institutionally recognized, are young, international professionals with varying qualifications and work experience who are employed as Teaching-and-Boarding Assistants at these sites. The analysis presented here explores how staffing issues, Brexit, and the ideological underpinnings of linguistic immersion serve to de-skill these mobile workers and render their position in these settings as that of precarious language workers or gap year students.

Finally, Neriko Musha Doerr brings the book to a powerful close in her afterward. Doerr highlights how all the participants in language-motivated voluntourism featured in this book—volunteer language teachers, hopeful language learners working without pay or for menial wages, volunteer interns, and gap year language assistants—make do with the "Wages of Global Experience" as compensation for their labor. In accepting this wage, volunteers are vulnerable to exploitation at the same time that "global experience," itself, is increasingly commodified.

Doerr's final comments leave us, however, with a nuanced pathway forward. As a "spark to chase and investigate further" (Doerr, this volume), Doerr proposes that what she calls "neo-immersion" could be a useful language learning method in many of the voluntourism contexts presented here. Dispelling with the notion of standardized languages that need to be learned in traditional classrooms, particular nation-states, or through formal instruction, the neo-immersion method involves "comfortable speakers (formerly known as native speakers)" working one-on-one with language learners and sharing their idiolects through casual conversations and interactions. Participants in these dyads are also compensated financially for their labor. And while it might be a stretch to think that neo-immersion will catch on widely anytime soon—particularly in language-motivated voluntourism contexts—Doerr's idea illustrates the mutuality, equitable nature, and fair(er) political-economic relations that critical (linguistic) projects are ultimately seeking.

References

Adichie, C. N. (2009). The danger of a single story. Talk presented at the TEDGlobal 2009 Conference in Oxford. Retrieved March 3, 2023, from https://www.ted.com/talks/chimamanda_ngozi_adichie_the_danger_of_a_single_story/c#t-1101088

Alim, H. S. (2005). Critical language awareness in the United States: Revisiting issues and revising pedagogies in a resegregated society. *Educational Researcher, 34*(7), 24–31. https://doi.org/10.3102/0013189X034007024

Alim, H. S., Rickford, J. R., & Ball, A. F. (2016). *Raciolinguistics: How language shapes our ideas about race.* Oxford University Press. https://doi.org/10.1093/acprof:oso/9780190625696.001.0001

Appadurai, A. (1996). *Modernity at large: Cultural dimensions of globalization.* University of Minnesota Press.

Bernstein, J., & Woosnam, K. M. (2019). Same same but different: Distinguishing what it means to teach English as a foreign language within the context of volunteer tourism. *Tourism Management, 72*(3), 427–436. https://doi.org/10.1016/j.tourman.2019.01.010

Blommaert, J. (2001). Context is/as critique. *Critique of Anthropology, 21*(1), 13–32. https://doi.org/10.1177/0308275X0102100102

Blommaert, J. (2009). Language, asylum, and the national order. *Current Anthropology, 50*(4), 415–441. https://doi.org/10.1086/600131

Blommaert, J. (2010). *The sociolinguistics of globalization.* Cambridge University Press. https://doi.org/10.1017/CBO9780511845307

Boutet, J. (2001). La part langagière du travail: Bilan évolution. *Langage et société, 98*(4), 17–42. https://doi.org/10.3917/ls.098.0017

Boutet, J. (2021). Des vocabulaires de métiers à la part langagière du travail. *Langues, Cultures et Sociétés, 7*(1), 22–31. https://doi.org/10.48384/IMIST.PRSM/lcs-v7i1.26764

Boutet, J., & Heller, M. (2007). Enjeux sociaux de la sociolinguistique: Pour une sociolinguistique critique. *Langage et société, 121–122*(3), 305–318. https://doi.org/10.3917/ls.121.0305

Brennan, C. (2014). Precarious working tourists: Working holiday makers in Australia. *Qualitative Sociology Review, 10*(3), 94–114. https://doi.org/10.18778/1733-8077.10.3.05

Brennan, C. (2018). Northern lights instead of workers' rights: Volunteer working tourists in Finnish Lapland. *Nordic Journal of Working Life Studies, 8*(2), 43–61. https://doi.org/10.18291/njwls.v8i2.106154

Browning, P., Highet, K., Azada-Palacios, R., Douek, T., Yue Gong, E., & Sunyol, A. (2022). Conspiring to decolonise language teaching and learning: Reflections and reactions from a reading group. *London Review of Education, 20*(1). https://uclpress.scienceopen.com/hosted-document?doi=10.14324/LRE.20.1.42

Bunce, P. (2016). Voluntary overseas English language teaching: A myopic, altruistic Hydra. In P. Bunce, R. Phillipson, V. Rapatahana, & R. Tupas (Eds.), *Why English? Confronting the Hydra* (pp. 106–117). Multilingual Matters. https://doi.org/10.21832/9781783095858-012

Bunten, A. C. (2008). Sharing culture or selling out? Developing the commodified persona in the heritage industry. *American Ethnologist, 35*(3), 380–395. https://doi.org/10.1111/j.1548-1425.2008.00041.x

Burrai, E., Mostafanezhad, M., & Hannam, K. (2017). Moral assemblages of volunteer tourism development in Cusco, Peru. *Tourism Geographies, 19*(3), 362–377. https://doi.org/10.1080/14616688.2016.1236145

Butcher, J. (2003). *The moralisation of tourism: Sun, sand and saving the world?* Routledge. https://doi.org/10.4324/9780203987025

Butcher, J. (2006). A response to building a decommodified research paradigm in tourism: The contribution of NGOs by Stephen Wearing, Matthew McDonald and Jess Ponting. *Journal of Sustainable Tourism,14*(3), 307–310. https://doi.org/10.1080/09669580608669061

Butcher, J., & Smith, P. (2010). 'Making a difference': Volunteer tourism and development. *Tourism Recreation Research, 35*(1), 27–36. https://doi.org/10.1080/02508281.2010.11081616

Butcher, J., & Smith, P. (2015). *Volunteer tourism: The lifestyle politics of international development*. Routledge. https://doi.org/10.4324/9781315796376

Callanan, M., & Thomas, S. (2005). Volunteer tourism: Deconstructing volunteer activities within a dynamic environment. In M. Novelli (Ed.), *Niche tourism: Contemporary issues and trends* (pp. 183–200). Elsevier.

Castillo Arredondo, M. I., Rodríguez Zapatero, M. I., Pérez Naranjo, L. M., & López-Guzmán, T. (2018). Motivations of educational tourists in non-English-speaking countries: The role of languages. *Journal of Travel & Tourism Marketing, 35*(4), 437–448. https://doi.org/10.1080/10548408.2017.1358238

Chen, L.-J. (2016). Intercultural interactions among different roles: A case study of an international volunteer tourism project in Shaanxi. *China. Current Issues in Tourism, 19*(5), 458–476. https://doi.org/10.1080/13683500.2015.1005581

Chun, J. J., & Han, J. J. (2015). Language travels and global aspirations of Korean youth. *Positions: Asia Critique, 23*(3), 565–593. https://doi.org/10.1215/10679847-3125913

Clarke, N. (2005). Detailing transnational lives of the middle: British working holiday makers in Australia. *Journal of Ethnic and Migration Studies, 31*(2), 307–322.

Codó, E. (2018). The intersection of global mobility, lifestyle and ELT work: A critical examination of language instructors' trajectories. *Journal of Language and Intercultural Communication, 18*(4), 436–450. https://doi.org/10.1080/14708477.2018.1482905

Codó, E., & McDaid, J. (2019). English language assistants in the 21st century: Nation-state soft power in the experience economy. *Language, Culture and Society, 1*(2), 219–243.

Cohen, E., & Cooper, R. L. (1986). Language and tourism. *Journal of Travel Research, 13*(4), 533–563. https://doi.org/10.1016/0160-7383(86)90002-2

Cohen, M. (1996). *Fashioning masculinity: National identity and language in the eighteenth century*. Routledge. https://doi.org/10.4324/9780203206102

Cole, T. (2016). *Known and strange things*. Random House.

Collins, F. L., & Shubin, S. (2015). Migrant times beyond the life course: The temporalities of foreign English teachers in South Korea. *Geoforum, 62*, 96–104. https://doi.org/10.1016/j.geoforum.2015.04.002

Dann, G. (1996). *The language of tourism: A sociolinguistic perspective*. CABI.

Dlaske, K. (2016). Shaping subjects of globalisation: At the intersection of voluntourism and the new economy. *Multilingua, 35*(4), 415–440. https://doi.org/10.1515/multi-2015-0002

Doerr, N. M. (Ed.) (2009). *The native speaker concept: Ethnographic investigations of native speaker effects*. Mouton De Gruyter.

Doerr, N. M., & Taïeb, H. D. (2020). *The romance of crossing borders: Studying and volunteering abroad*. Berghahn Books.

Drozdzewski, D. (2011). Language tourism in Poland. *Tourism Geographies, 13*(2), 165–186. https://doi.org/10.1080/14616688.2011.569569

Duchêne, A. (2009). Marketing, management and performance: Multilingualism as commodity in a tourism call centre. *Language Policy, 8*, 27–50. https://doi.org/10.1007/s10993-008-9115-6

Duchêne, A. (2019). Unequal language work(ers) in the business of words. In J. Thurlow (Ed.), *The business of words: Wordsmiths, linguists and other language workers* (pp. 23–35). Routledge.

Duchêne, A., & Daveluy, M. (2015). Présentation: Spéculations langagières: Négocier des ressources aux valeurs fluctuantes. *Anthropologie et Sociétés, 39*(3), 9–27. https://doi.org/10.7202/1034757ar

Duchêne, A., & Piller, I. (2011). Mehrsprachigkeit als Wirtschaftsgut: Sprachliche Ideologien und Praktiken in der Tourismusindustrie. In G. Kreis (Ed.), *Babylon Europa: Zur europäischen Sprachlandschaft* (pp. 135–157). Schwabe.

Duncan, T. (2007). *Working tourists: Identity formation in a leisure space.* [Unpublished doctoral dissertation]. University College London.

Duncan, T., Scott, D. G., & Baum, T. (2013). The mobilities of hospitality work: An exploration of issues and debates. *Annals of Tourism Research, 41*, 1–19. https://doi.org/10.1016/j.annals.2012.10.004

Everingham, P. (2018). Speaking Spanglish: Embodying linguistic (b)orderlands in volunteer tourism. *Emotion, Space, and Society, 27*, 68–74. https://doi.org/10.1016/j.emospa.2018.04.001

Fraser, N., & Honneth, A. (2003). *Redistribution or recognition? A political-philosophical exchange.* Verso.

Gao, S. (2019). *Aspiring to be global: Language and social change in a tourism village in China.* Multilingual Matters. https://doi.org/10.1515/9781788922760

Geertz, C. (1973). Thick description: Toward an interpretive theory of culture. In *The interpretation of culture: Selected essays* (pp. 3–30). Basic Books.

Germann Molz, J. (2021). *The world is our classroom: Extreme parenting and the rise of worldschooling.* New York University Press.

Giddens, A. (1994). *Beyond left and right: The future of radical politics.* Polity.

Goldoni, F. (2013). Students' immersion experiences in study abroad. *Foreign Language Annals, 46*(3), 359–376. https://doi.org/10.1111/flan.12047

Haslanger, S. (2021). Methods of social critique. In A. Siegetsleitner, A. Oberprantacher, M. Frick, & U. Metschl (Eds.), *Crisis and critique: Philosophical analysis and current events: Proceedings of the 42nd international Ludwig Wittgenstein symposium* (pp. 139–156). De Gruyter Mouton. https://doi.org/10.1515/9783110702255-010

Heller, M. (2001). Critique and sociolinguistic analysis of discourse. *Critique of Anthropology, 21*(2), 117–141. https://doi.org/10.1177/0308275X0102100201

Heller, M. (2002). *Éléments d'une sociolinguistique critique.* Didier.

Heller, M. (2010). The commodification of language. *Annual Review of Anthropology, 39*, 101–114. https://doi.org/10.1146/annurev.anthro.012809.104951

Heller, M., Jaworski, A., & Thurlow, C. (2014). Introduction: Sociolinguistics and tourism – Mobilities, markets, multilingualism. *Journal of Sociolinguistics, 18*(4), 425–458. https://doi.org/10.1111/josl.12091

Henry, J. (2018a). Pedagogy, possibility, and pipe dreams: Opportunities and challenges for radicalizing international volunteering. *Journal of Tourism & Cultural Change, 17*(6), 661–673. https://doi.org/10.1080/14766825.2018.1515215

Henry, J. (2018b). The unspeakable whiteness of volunteer tourism. *Annals of Tourism Research, 76*(C), 326–327. https://doi.org/10.1016/j.annals.2018.09.003

Henry, J. (2020). The cinematic pedagogies of underprepared teachers. *Teaching & Teacher Education, 89*, 1–9. https://doi.org/10.1016/j.tate.2019.102990

Henry, J., & Mostafanezhad, M. (2019). The geopolitics of volunteer tourism. In D. J. Timothy (Ed.), *Handbook of globalisation and tourism* (pp. 295–304). Edward Elgar Publishing.

Highet, K., & Schedel, L. S. (Eds.) (forthcoming). Language, work and social critique: In dialogue with the translated works of Josiane Boutet [Special issue]. International Journal of the Sociology of Language.

Highet, K., Makoni, S., Urciuoli, B., & Urla, J. (forthcoming). Critique. In A. Del Percio & M.-C. Flubacher (Eds.), *Critical sociolinguistic dialogues: The constitution of a field*. Bloomsbury.

Horner, J., & Dailey-O'Cain, J. (2019). *Multilingualism, (im)mobilities and spaces of belonging*. Multilingual Matters. https://doi.org/10.21832/9781788925051

Iglesias, M. (2016). The language tourism market system: Conceptualising language tourism. *International Journal of Scientific Management and Tourism, 2*(1), 25–40.

Iglesias, M. (2022). Language tourism. In M. Novelli, J. Cheer, C. Dolezal, A. Jones, & C. Milano (Eds.), *Handbook of niche tourism* (pp. 218–231). Edward Elgar Publishing.

Ismail, N. (2015). *Ungehörte Stimmen: Repräsentation und Wissensproduktion aus postkolonial-feministischer Perspektive*. Löcker.

Jack, G., & Phipps, A. (2012). Tourism. In J. Jackson (Ed.), *The Routledge handbook of language and intercultural communication* (pp. 537–550). Routledge.

Jakubiak, C. (2012). "English for the global": Discourses in/of English-language voluntourism. *International Journal of Qualitative Studies in Education, 25*(4), 435–451. https://doi.org/10.1080/09518398.2012.673029

Jakubiak, C. (2016a). Ambiguous aims: English-language voluntourism as development. *Journal of Language, Identity & Education, 15*(4), 245–258. https://doi.org/10.1080/15348458.2016.1195270

Jakubiak, C. (2016b). A pedagogy of enthusiasm: A critical view of English-language voluntourism. In J. A. Álvarez Valencia, C. Amanti, S. Keyl, & E. Mackinney (Eds.), *Critical views on teaching and learning English around the globe: Qualitative research approaches* (pp. 193–209). Information Age Publishing.

Jakubiak, C. (2020). "English is out there—You have to get with the program": Linguistic instrumentalization, global citizenship education, and English-language voluntourism. *Anthropology & Education Quarterly, 51*(2), 212–232. https://doi.org/10.1111/aeq.12332

Jakubiak, C. (2021). The possibility of critical language awareness through volunteer English teaching abroad. In K. Swalwell & D. Spikes (Eds.), *Anti-oppressive education in "elite" schools: Promising practices and cautionary tales from the field* (pp. 109–121). Teachers College Press.

Jang I. C. (2017). *Consuming global language and culture: South Korean youth in English study abroad.* [Doctoral dissertation, University of Toronto]. Tspace Repository. https://tspace.library.utoronto.ca/bitstream/1807/79006/3/Jang_In_Chull_201706_PhD_thesis.pdf

Jaworski, A., & Pritchard, A. (2005). *Discourse, communication, and tourism.* Channel View Publications. https://doi.org/10.21832/9781845410216

Jaworski, A., Thurlow, C., & Heller, M. (2014). Sociolinguistics and tourism [Special issue]. *Journal of Sociolinguistics, 18*(4), 422–566.

Jenks, C. J., & Lee, J. W. (2020). Native speaker saviorism: A racialized teaching ideology. *Critical Inquiry in Language Studies, 17*(3), 186–205. https://doi.org/10.1080/15427587.2019.1664904

Keese, J. R. (2011). The geography of volunteer tourism: Place matters. *Tourism Geographies, 13*(2), 257–279. https://doi.org/10.1080/14616688.2011.567293

Kinginger, C. (2004). Alice doesn't live here anymore: Foreign language learning and identity reconstruction. In A. Pavlenko & A. Blackledge (Eds.), *Negotiating identities in multilingual contexts* (pp. 219–242). Channel View Publications. https://doi.org/10.21832/9781853596483-010

Knott, A. (2018). Guests on the Aegean: Interactions between migrants and volunteers at Europe's southern border. *Mobilities, 13*(3), 349–366. https://doi.org/10.1080/17450101.2017.1368896

Laesser, C., Beritelli, P., & Heer, S. (2014). Different native languages as proxy for cultural differences in travel behaviour: Insights from multilingual Switzerland. *International Journal of Culture, Tourism and Hospitality Research, 8*(2), 140–152. https://doi.org/10.1108/IJCTHR-02-2014-0010

Lee, J. S. (2011). Globalization and language education: English village in South Korea. *Language Research, 47*(1), 123–149.

Lo, A. S., & Lee, C. Y. S. (2011). Motivations and perceived value of volunteer tourists from Hong Kong. *Tourism Management, 32*, 326–334. https://doi.org/10.1016/j.tourman.2010.03.002

Lyons, K., & Wearing, S. (2012). Reflections on the ambiguous intersections between volunteering and tourism. *Leisure Sciences, 34*(1), 88–93. https://doi.org/10.1080/01490400.2012.633858

Lyons, K. D., & Wearing, S. (2008). All for a good cause? The blurred boundaries of volunteering and tourism. In K. D. Lyons & S. Wearing (Eds.), *Journeys of discovery in volunteer tourism: International case study perspectives* (pp. 147–154). CABI.

Marty Crettenand, S. (2016). Le rôle de la langue dans le projet de migration de jeunes femmes suisses alémaniques vers la Suisse romande. In M. Matthey & A. Millet (Eds.), *Hétérogénéité et changement: Perspectives sociolinguistiques. Actes du 2ème congrès du réseau francophone de sociolinguistique, Grenoble, 10-11 juin 2015* (pp. 109–116) EME Editions.

McBride, A. M., & Sherraden, M. (2007). *Civic service worldwide: Impacts and inquiry*. Routledge.

McGehee, N. G., & Santos, C. A. (2005). Social change, discourse, and volunteer tourism. *Annals of Tourism Research, 32*(3), 760–779. https://doi.org/10.1016/j.annals.2004.12.002

Moïse, C., & Wilson, A. (Eds.) (2023). La part langagière du tourisme [Special issue]. *Langage et Societe, 178*(1), 225 pages.

Mostafanezhad, M., & Hannam, K. (2014). *Moral encounters in tourism*. Routledge. https://doi.org/10.4324/9781315596068

Motobayashi, K. (2015). *Language teaching as foreign policy: Japanese language teachers in Japan's international cooperation volunteer program*. [Doctoral dissertation. University of Toronto]. TSpace Repository. https://hdl.handle.net/1807/69470

Mustonen, P. (2006). Volunteer tourism: Postmodern pilgrimage? *Journal of Tourism and Cultural Change, 3*, 160–177. https://doi.org/10.1080/14766820608668493

Muth, S. (2018). "The ideal Russian speaker is no Russian": Language commodification and its limits in medical tourism to Switzerland. *Language*

Policy, 17(2), 217–237. https://link.springer.com/article/10.1007/s10993-017-9434-6

Nye, J. S., Jr. (2004). Soft power and American foreign policy. *Political Science Quarterly, 119*(2), 255–270. https://doi.org/10.2307/20202345

O'Rourke, B., & DePalma, R. (2017). Language-learning holidays: What motivates people to learn a minority language? *International Journal of Multilingualism, 14*(4), 1–18. https://doi.org/10.1080/14790718.2016.1184667

Palacios, C. M. (2010). Volunteer tourism, development, and education in a postcolonial world: Conceiving global connections beyond aid. *Journal of Sustainable Tourism, 18*(7), 861–878. https://doi.org/10.1080/09669581003782739

Park, J. H. (2018). Cultural implications of international volunteer tourism: U.S. students' experiences in Cameroon. *Tourism Geographies, 20*(1), 144–162. https://doi.org/10.1080/14616688.2017.1387810

Park, J. S.-Y. (2009). *The local construction of a global language: Ideologies of English in South Korea*. De Gruyter Mouton. https://doi.org/10.1515/9783110214079

Park, J. S.-Y., & Gao, S. (2015). Space and language learning under the neoliberal economy. *L2 Journal, 7*(3), 78–96. https://doi.org/10.5070/L27323514

Parreñas, J. S. (2016). The materiality of intimacy: Rethinking "ethical capitalism" through embodied encounters with animals in Southeast Asia. *Positions: Asia Critique, 24*(1), 97–127. https://doi.org/10.1215/10679847-3320065

Pennycook, A. (2020). Translingual entanglements of English. *World Englishes, 39*(2), 222–235. https://doi.org/10.1111/weng.12456

Pennycook, A., & Makoni, S. (2019). *Innovations and challenges in applied linguistics from the Global South*. London: Routledge. https://doi.org/10.4324/9780429489396

Petit Cahill, K. (2020). Creating places through language rules: A historical and ethnographic perspective on the 'rule of Irish'. *Journal of Sociolinguistics, 24*(2), 228–244. https://doi.org/10.1111/josl.12365

Phillipson, R. (1992). *Linguistic imperialism*. Oxford University Press.

Phipps, A. (2006). *Learning the arts of linguistic survival: Languaging, tourism, life*. Channel View Publications. https://doi.org/10.21832/9781845410551

Pietikäinen, S., & Kelly-Holmes, H. (2011). The local political economy of languages in a Sámi tourism destination: Authenticity and mobility in the labelling of souvenirs. *Journal of Sociolinguistics, 15*(3), 323–346. https://doi.org/10.1111/j.1467-9841.2011.00489.x

Piller, I. (2017). *Intercultural communication: A critical introduction* (2nd ed.). Edinburgh University Press.

Prince, S. (2017). Working towards sincere encounters in volunteer tourism: An ethnographic examination of key management issues at a Nordic eco-village. *Journal of Sustainable Tourism, 25*(11), 1617–1632. https://doi.org/10.1080/09669582.2017.1297450

Projects Abroad. (n.d.). *What is voluntourism?* Retrieved April 19, 2023, from https://www.projects-abroad.org/voluntourism/#:~:text=A%20combination%20of%20volunteering%20and,a%20different%20country%20and%20culture

Rey, J., Bolay, M., & Gez, Y. N. (2020). Precarious privilege: Personal debt, lifestyle aspirations and mobility among international school teachers. *Globalisation, Societies and Education, 18*(4), 361–373. https://doi.org/10.1080/14767724.2020.1732193

Rice, K. (2010). 'Working on holiday': Relationships between tourism and work among young Canadians in Edinburgh. *Anthropology in Action, 17*(1), 30–40. https://doi.org/10.3167/aia.2010.170104

Salazar, N. B. (2005). Tourism and glocalization: "Local" tour guiding. *Annals of Tourism Research, 32*(3), 628–646. https://doi.org/10.1016/j.annals.2004.10.012

Schedel, L. S. (2018a). Negotiating the boundaries and value of bilingual skills in a recruitment process at the language border. *Bulletin Suisse de Linguistique Appliquée, 108*, 75–90. https://www.e-periodica.ch/cntmng?pid=val-003%3A2018%3A0%3A%3A321

Schedel, L. S. (2018b). Turning local bilingualism into a touristic experience. *Language Policy, 17*(2), 137–155. https://doi.org/10.1007/s10993-017-9437-3

Schedel, L. S. (2022). The price of immersion: Language learners as a cheap workforce in Malta's voluntourism industry. *Multilingua, 41*(2), 181–200. https://doi.org/10.1515/multi-2020-0178

Schedel, L. S. (2023). Exploiter le plurilinguisme pour vendre une expérience 'monolingue': Le cas du tourisme linguistique à Malte. *Langage et société, 178*(1), 75–94. https://doi.org/10.3917/ls.178.0066

Schwarz, K. C. (2018). Volunteer tourism and the intratourist gaze. *Tourism Recreation Research, 43*(2), 186–196. https://doi.org/10.1080/02508281.2017.1399543

Schwarz, K. C., & Richey, L. A. (2019). Humanitarian humor, digilantism, and the dilemmas of representing volunteer tourism on social media. *New Media & Society, 21*(9), 1928–1946. https://doi.org/10.1177/1461444819834509

Scott, J. C. (1998). *Seeing like a state: How certain schemes to improve the human condition have failed.* Yale University Press. https://www.jstor.org/stable/j.ctt1nq3vk

Sharma, B. K., & Gao, S. (Eds.). (2022). *Language and intercultural communication in tourism: Critical perspectives*. Routledge. https://doi.org/10.4324/9781003088028

Sheller, M. (2014). The new mobilities paradigm for a live sociology. *Current Sociology, 62*(6), 789–811. https://doi.org/10.1177/0011392114533211

Simpson, K. (2004). 'Doing development': The gap year, volunteer-tourists and a popular practice of development. *Journal of International Development, 16*(5), 681–692. https://doi.org/10.1002/jid.1120

Simpson, K. (2005). Dropping out or signing up? The professionalisation of youth travel. *Antipode, 37*(3), 447–469. https://doi.org/10.1111/j.0066-4812.2005.00506.x

Sin, H. S. (2009). Volunteer tourism—"Involve me and I will learn?". *Annals of Tourism Research, 36*(3), 480–501. https://doi.org/10.1016/j.annals.2009.03.001

Song, J. (2011). Globalization, children's study abroad, and transnationalism as an emerging context for language learning: A new task for language teacher education. *TESOL Quarterly, 45*(4), 749–758. https://www.jstor.org/stable/41307664

Stainton, H. (2016). A segmented volunteer tourism industry. *Annals of Tourism Research, 61*(C), 256–258. https://doi.org/10.1016/j.annals.2016.09.011

Stainton, H. (2018). TEFL tourism: The tourist who teaches. *Tourism Geographies, 20*(1), 127–143. https://doi.org/10.1080/14616688.2017.1298151

Stainton, H. (2019). *TEFL tourism: Principles, commodification and the sustainability of teaching English as a foreign language*. Routledge.

Stanley, P. (2013). *A critical ethnography of 'Westerners' teaching English in China: Shanghaied in Shanghai*. Routledge. https://doi.org/10.4324/9780203078051

Stanley, P. (2017). *A critical auto/ethnography of learning Spanish: Intercultural competence on the gringo trail?* Routledge. https://doi.org/10.4324/9781315708232

Takahashi, K. (2013). *Language learning, gender and desire: Japanese women on the move*. Multilingual Matters. https://doi.org/10.21832/9781847698551

Thomas-Maude, J., McLennan, S., & Walters, V. (2021). Cultural exchange during English-language voluntourism (EVT) in Lima, Peru: A postcolonial analysis. *Journal of Language, Identity & Education, First online*. https://doi.org/10.1080/15348458.2021.1991802

Thurlow, C., & Jaworski, A. (2010). *Tourism discourse: Language and global mobility*. Palgrave Macmillan.

Thurlow, C., & Jaworski, A. (2011). Tourism discourse: Languages and banal globalization. *Applied Linguistics Review, 2*, 285–312. https://doi.org/10.1515/9783110239331.285
Tiessen, R. (2018). *Learning and volunteering abroad for development: Unpacking host organization and volunteer rationales*. Routledge.
Tosi, A. (2020). *Language and the Grand Tour: Linguistic experiences of travelling in early modern Europe*. Cambridge University Press. https://doi.org/10.1017/9781108766364
TRAM. (2008). *Volunteer tourism: A global analysis*. ATLAS.
Tupas, R. (2020). Decentering language: Displacing Englishes from the study of Englishes. *Critical Inquiry in Language Studies, 17*(3), 228–245. https://doi.org/10.1080/15427587.2019.1641097
Urla, J. (2021). Una crítica generativa de la gubernamentalidad lingüística (L. Alonso & J. Alvis, Eds.). *Anuario de Glotopolítica, 4*. https://glotopolitica.com/aglo-4/urla/
Vrasti, W. (2013). *Volunteer tourism in the Global South: Giving back in neoliberal times*. Routledge. https://doi.org/10.4324/9780203104453
Wearing, S. (2001). *Volunteer tourism: Experiences that make a difference*. CABI.
Wearing, S., Deville, A., & Lyons, K. (2008). The volunteer's journey through leisure into the self. In K. D. Lyons & S. Wearing (Eds.), *Journeys of discovery in volunteer tourism: International case study perspectives* (pp. 63–71). CABI.
Wearing, S., McDonald, M., & Ponting, J. (2005). Building a decommodified research paradigm in tourism: The contribution of NGOs. *Journal of Sustainable Tourism, 13*(5), 424–439. https://doi.org/10.1080/09669580508668571
Wearing, S., Mostafanezhad, M., Nguyen, N., Nguyen, T. H. T., & McDonald, M. (2018). "Poor children on Tinder" and their Barbie sSaviours: Towards a feminist political economy of volunteer tourism. *Leisure Studies, 37*(5), 500–515. https://doi.org/10.1080/02614367.2018.1504979
Wilson, A. (2016). "Parler touriste" : La mobilité de langues, de locuteurs et de normes dans les interactions exolingues en milieu touristique. In M. Matthey & A. Millet (Eds.), *Hétérogénéité et changement: Perspectives sociolinguistiques. Actes du 2ème congrès du réseau francophone de sociolinguistique, Grenoble, 10-11 juin 2015* (pp. 193–200). EME Editions.
Wilson, A. (2018). Adapting English for the specific purpose of tourism: A study of communication strategies in face-to-face encounters in a French tourist office. *ASp, 73*, 53–73. https://journals.openedition.org/asp/5118#quotation

Xiang, B., & Lindquist, J. (2014). Migration infrastructure. *International Migration Review, 48*(1_suppl), 122–148. https://doi.org/10.1111/imre.12141

Yarymowich, M. (2004). "Language tourism" in Canada: A mixed discourse. In F. Baider, M. Burger, & D. Goutsos (Eds.), *La communication touristique : Approches discursives de l'identité et de l'alterité* (pp. 257–273). L'Harmattan.

Yoon, K. (2014). The racialised mobility of transnational working holidays. *Identities: Global Studies in Culture and Power, 21*(5), 586–603. https://doi.org/10.1080/1070289X.2014.909815

Part I

Language-Motivated Voluntourism in Contexts of Leisure and Holiday Travel

2

Immersion as Language Ideology and Other Discourses in English-Language Voluntourism

Cori Jakubiak

Introduction

The turn of the twenty-first century marked a swell of interest in volunteer tourism, which Stephen Wearing defined in 2001 as "volunteer[ing] in an organized way to undertake holidays that might involve aiding or alleviating the material poverty of some groups in society, the restoration of certain environments, or research into aspects of society or environment" (p. 1). While the meaning of *volunteer tourism* is now expanding to include a variety of precarious leisure/work mobilities that involve voluntary labor and travel (see Schedel, 2021, and this volume), trade book titles such as *Volunteer Vacations: Short-Term Adventures that will Benefit You and Others* (McMillion et al., 2012) and *The 100 Best Volunteer Vacations to Enrich Your Life* (Grout, 2009) suggest that humanitarian-oriented volunteer tourism is likely to be with us for some time. Indeed,

C. Jakubiak (✉)
Grinnell College, Grinnell, IA, USA
e-mail: jakubiak@grinnell.edu

L. S. Schedel, C. Jakubiak (eds.), *Voluntourism and Language Learning/Teaching*, Palgrave Advances in Language and Linguistics,
https://doi.org/10.1007/978-3-031-40813-7_2

opportunities to combine a traditional holiday with international volunteering abound, with novel forms of volunteer tourism such as "voluncruising" (Watkins, 2013) and even "honeyteering" (Knight, 2023) emerging at the leisure/service nexus. As the U.K.-based organization, Volunteer Latin America (2022), advertises on its website, "From volunteering with wildlife to working in an orphanage, from construction to teaching English, there are thousands of organisations all over the world eager for help."

Dangling modifiers aside, Volunteer Latin America's (2022) marketing statement is perplexing in numerous ways. While it provides an accurate summary of the kinds of activities in which volunteer tourists engage—that is, assisting with wildlife care, working in orphanages, doing light construction work, teaching English—it collapses all these activities together in the name of "help." This rhetorical move raises a series of questions, particularly from a critical sociolinguistics standpoint. What is at play when short-term English language teaching (ELT) is framed as humanitarian assistance, particularly in settings where English is neither a dominant nor community language? In what social context(s) can short-term English language teaching be equivalent to providing vulnerable people with basic care? This chapter grapples with these and related questions.

In the pages that follow, I examine how language education is discussed within *English-language voluntourism*, or short-term, volunteer English language teaching in the Global South (see also Jakubiak, 2016, 2020). Like other forms of volunteer tourism in the Wearing (2001) tradition, English-language voluntourism is characterized as unskilled, humanitarian work. Teachers in English-language voluntourism programs generally need not be conversant in host community language(s) nor possess professional educator credentials. Instead, non-governmental organizations (NGOs) and commercial sponsors offer that volunteers' enthusiasm and native English-speaking abilities are sufficient qualifications to teach (see also Bunce, 2016, for a critique of these minimal qualifications).

Drawing upon data collected through a discourse analysis of volunteer tourism promotional materials and qualitative interviews with in-service and former volunteers, this chapter traces the language ideologies that adhere to English-language voluntourism. These language ideologies, assembled together, construct short-term, volunteer English language teaching in the Global South as a necessary and benevolent intervention. However, not all of my study participants took up these language ideologies uncritically. As my data analysis will show, some former volunteers take issue with English-language voluntourism's dominant discourses—particularly its stress on what I call "native English immersion as pedagogy." This discourse does not always stand up to scrutiny in real-time, real-life classrooms, and the dissonance between promotional rhetoric and volunteers' actual experiences leads some former volunteer teachers to contest dominant volunteer tourism discourses with humility.

English as Development Within English-Language Voluntourism

Central to my analysis of English-language voluntourism discourse is Roslyn Appleby et al.'s (2002) distinction between *English for development* and *English as development* ideologies. In the former ideology, English language skills are described as helping people to participate in existent development projects. In the latter, English language learning itself is the development aim (see Appleby et al., 2002). I have argued that *English as development* ideology predominates in English-language voluntourism (see Jakubiak, 2016, 2020). However, the various, even conflicting, ways in which English-language voluntourism discourse operationalizes *English as development* ideology illuminate facets of the neoliberal present and the role of English within it.

As I detail in this chapter, *English as development* within English-language voluntourism discourse reflects four key positions. First, and most infrequently, it indexes broad-scale national development. I refer to this discourse as *English for modernist development*, as it aims to foster widespread transformation through centrally planned, technicist projects (cf. Kapoor, 2004). Second, and deployed much more frequently within English-language voluntourism discourse, *English as development* is tied to personal development, or the construction of particular kinds of people: entrepreneurial, flexible, life-long learner-subjects whose plurilingualism services neoliberal capitalism (see De Costa et al., 2019). I term this second discourse *English for entrepreneurial subjectivity*, and I have written about it extensively elsewhere (see Jakubiak, 2020). In its third iteration, *English as development* within English-language voluntourism ties uneven English language education to persistent and longstanding inequality within a nation-state. What I call *English for addressing inequality* discourse suggests that more equitable English language education can mitigate other inter-group disparities. Finally, *English as development* within English-language voluntourism seeks to remedy existent English language education in the Global South. In what I term *native English immersion as pedagogy* discourse, local teachers' English language uses are cast as prescriptively incorrect and their pedagogical methods as unsound. This final discourse suggests that prestige-variety, native English language speakers are urgently needed throughout the Global South to model prescriptive language use and foster communicative language teaching. It is on these last two versions of *English as development* discourse that this chapter primarily focuses.

I lay out my argument in the text of this chapter as follows. First, I situate English-language voluntourism within a body of related literature, arguing that segmented volunteer tourism studies (Stainton, 2016) provide an important corrective to a longstanding tradition of volunteer tourism being analyzed as singular phenomenon (see also Jakubiak & Schedel, this volume). Then, after discussing my study's methods and glossing the first two *English as development* discourses described above, my focus shifts to a more thorough explication of the last two discourses: *English for addressing inequality* and *native English immersion as pedagogy*. I suggest that both *English for addressing inequality* and *native English immersion as pedagogy* discourses draw heavily upon what Christopher J. Jenks and Jerry Won Lee (2020) term *native speaker saviorism* to frame English-language voluntourism in the Global South as a beneficent and necessary intervention. Native speaker saviorism draws attention to the ways in which a "White savior imperative" (Jenks & Lee, 2020, p. 189) of noblesse oblige adheres to the *native speaker fallacy* (Phillipson, 1992), or the idea that any native speaker of prestige-variety English is innately well-suited to teach it. Native speaker saviorism thus extends and refines the native speaker fallacy by explicitly tracing its racialized and affective aspects.

Finally, I turn to discrepant data to highlight how some of my study participants explicitly rejected *native English immersion as pedagogy* discourse. I suggest that because the claims of *native English immersion as pedagogy* discourse can actually be field-tested, the discourse rubs up uncomfortably against the realities of some volunteers' experiences. Certain volunteers' own instructional ineffectiveness and interdependency on multilingual hosts tempered the hubris associated with *native English immersion as pedagogy* discourse, leading them to develop more *critical language awareness* (Alim, 2005) by having taught English abroad (see also Jakubiak, 2021).

Beyond Good or Bad: Studies of Volunteer Tourism in Specific Sectors

Public and academic critiques of volunteer tourism have risen rapidly in recent years. While early studies celebrated volunteer tourism as a decommodified, alternative travel form that harnessed youths' civic engagement aspirations (e.g., Wearing, 2001), recent studies have taken a more critical turn. Humanitarian-oriented volunteer tourism is now often seen as a part of a "White Savior Industrial Complex" (Cole, 2016) in which middle-class, well-meaning, often White people pay to save vulnerable people in the Global South through un-reflexive and paternalistic interventions (Henry, 2020). Volunteer tourism has been disparaged for prioritizing volunteers' needs over those of host communities (Knott, 2018); teaching volunteers little about structural poverty (Crossley, 2012; Vrasti, 2013); and exploiting vulnerable peoples' images for social media purposes (Schwarz & Richey, 2019). To be sure, whether volunteer tourism "does more harm than good" has become something of a mantra in public discussions (see Jakubiak & Schedel, this volume).

A parallel scholarly development has been a rise in studies that examine particular *kinds of* volunteer tourism. Rather than wrestling over whether volunteer tourism is good or bad or offering suggestions on how these programs might be improved, studies of segmented volunteer tourism (following Stainton's 2016 call) offer insights on wider social phenomena. That is, careful analyses of unique volunteer tourism *forms* are shedding nuanced light on broader political issues and the contours of structural inequality. Studies of volunteer tourism in refugee arrival and encampment settings, for example, have contributed numerous insights to im/migration theorizing. Knott's (2018) multi-sited ethnography of volunteer tourism on the Greek islands of Lesvos and Chios during the 2015 refugee crisis at Europe's southern border describes how refugee "hungry" volunteer tourists, "who had anticipated saving, interacting with, and caring for arrivals" (p. 351), rushed beaches during asylum seekers' landings and physically grabbed children off of rubber boats. These interventions separated children from their parents and exacerbated asylum seekers' trauma. Still, the presence of well-meaning but ineffective volunteer tourists was legally authorized in Greece while that of asylum seekers was not. Observing this paradox, Knott (2018) writes:

> [V]olunteers are able to come from all over the world for a few hundred dollars by safe and comfortable modes of transport to 'help' asylum seekers, while the latter pay thousands of dollars to cross a narrow stretch of sea in an overcrowded and unseaworthy boat. The main difference between them is the possession of the right kinds of documentation. (p. 350)

Examining volunteer tourism in the context of refugee arrival and encampment contexts highlights the irrationalities of contemporary mobility assemblages. These assemblages sanction volunteer tourists' international movements while vilifying those of others (see also Doerr, this volume).

Studies of volunteer tourism in animal sanctuary contexts also underscore key features and problematics of the present. Researchers note that volunteer tourists' interventions in Global South animal sanctuaries can entrench globally reaching gendered and racialized labor hierarchies (Collard, 2020; Parreñas, 2012). In Rheana Juno Salazar Parreñas' (2012) ethnography of volunteer tourism in a Malaysian wildlife center, for example, female, White, middle-class volunteer tourists sought the affective experiences of hard physical labor and safe interactions with orangutans as a reprieve from the alienation of their Global North desk jobs. By contrast, underpaid, local Malaysian men held permanent jobs doing the most dangerous, physical, and dirty work at the sanctuary: cleaning and maintaining animal confinements. One of the full-time sanctuary workers in Parreñas' study even died after contracting a zoonotic virus from extended close contact with a captive orangutan—an experience from which visiting, White, female volunteer tourists are protected.

To be sure, multispecies encounters among volunteer tourists, full-time workers, and non-human animals in Global South sanctuary settings can reinscribe colonial-era, gendered, and racialized labor hierarchies that prioritize the safety and interests of White women over colonized people and captive animals. Commenting on this phenomenon, Parreñas (2012) writes:

> The participation and vulnerability of these particular subjects [in volunteer tourism] is unequally distributed between those who see such a project as a vacation, those whose employment is defined by it, and those displaced animals that have … nowhere else to go. (p. 683)

In other words, actors in volunteer tourism constellations play various, unequal, gendered, and racialized roles—helper and helped—in which risk is disproportionately borne by marginalized groups.

Studies of segmented volunteer tourism, then, offer fresh insights on wider social issues such as immigration policy, labor hierarchies, and animal rights. Yet, research on English-language voluntourism, specifically, has remained largely centered on whether the practice is good or bad (Bunce, 2016); whether English-language voluntourism constitutes a differentiated tourism form at all (Bernstein & Woosnam, 2019); and how these programs might be improved through more professionalized language teaching (Carpenter, 2015). This chapter thus contributes to growing body of work that elucidates key features of the contemporary social world by examining particular volunteer tourism forms—in this case, language teaching (see also Thomas-Maude et al., 2021, advance online publication).

Methods

Data for this study were collected between 2007 and 2014, when I conducted a multi-sited ethnography (Marcus, 1995) of English-language voluntourism in three different contexts. I taught English as a volunteer tourist in a public junior high for two months in a northern Costa Rica village; worked as a volunteer program assistant for one month in the northeastern U.S. offices of the NGO that had sponsored my Costa Rica program; and volunteered for two weeks in a Romanian intentional community, where English was taught by NGO-sponsored volunteers on an ad hoc basis. In these participant-observation contexts, I conducted and collected over 40[1] qualitative interviews with in-service volunteers and NGO staff workers. I also conducted and collected 20 interviews with former volunteers, whom I contacted through the NGO with which I had volunteered in Costa Rica. Three of these interviews with former volunteers can be classified as *return interviews* (Tobin & Hayashi, 2017), as these participants were members of my own Costa Rica cohort and talked with me while volunteering as well as retrospectively. Additionally, I conducted a discourse analysis of over 20 different volunteer tourism sponsors' promotional websites, identifying the language ideologies that they used in their descriptions of English-language voluntourism's purposes and goals.

My data analysis used a grounded theory approach, following Kathy Charmaz (2006). I coded interview transcripts and text from promotional websites according to four emergent themes. These themes, which I label as discourses here, reveal the language ideologies that circulated in volunteers' talk and in promotional websites' descriptions of English-language voluntourism. It is to a discussion of these findings that I now turn.

[1] An undergraduate researcher assisted me in collecting interviews in Romania. However, we ultimately decided against using some of these interviews due to concerns about interviewees' discomfort.

Findings

English for Modernist Development

On English-language voluntourism sponsors' websites and in stakeholders' talk, *English as development* ideology occasionally indexes what Ilan Kapoor (2004) terms the modernist version of development. Here, development refers to structural interventions designed to promote national-level social or economic transformation. This view of development has "a close affinity with teleological views of history, science, and progress in the West" (Peet & Watts, 1993, p. 232), and it manifests through centrally planned projects such as agricultural reforms and hydroelectric dams, the stated purpose of which is to strengthen a country's physical and social infrastructure (cf. Quesada, 2014).

What I refer to as *English for modernist development* discourse includes attention to a nation-state's linguistic resources, which are suggested to build a country's technical expertise, increase its access to the scientific community, and foster economic competitiveness (Kamwangamalu, 2008; see also Choi, this volume). In this frame, English language capacity among a national populace is tied to the development of human capital, a greater amount of which is argued to "enhance [peoples'] capacities as producers and consumers" (Schultz cited in Erling & Seargeant, 2013, p. 11) and increase a county's overall economic growth.

English-language voluntourism promotional literature occasionally employs *English for modernist development* discourse. In illustration, the NGO, New Hope Volunteers (2015) warrants its English-language voluntourism program in China by stating:

> China is growing incessantly. With this growth, [*sic*] comes constant beneficial development for the country. China's industrial and financial involvement in the international arena means that the English language is becoming a must-have skill for the people.

Implied here is that without a high level of English language skills across the country, China's international business interests will be truncated. *English for modernist development* discourse suggests a causal relationship between English language fluency among a nation-state's population and a nation-state's economic position. The discourse posits that a county's commercial success is contingent upon a high percentage of its citizens knowing English.

Scholarship refutes the claim, however, that increased English language capacity among the people of a nation-state will strengthen a county's economic standing. English language knowledge among a country's population is correlated to other variables such as levels of higher education, access to employment, and existent national infrastructure; it is not a causal factor in a nation-state's economic position. As language economists Jean-Louis Arcand & François Grin (2013) report:

> [W]idespread competence in a dominant language such as English is in no matter associated with a higher level of economic development, when the latter is measured by its most common incarnation of GDP per capita. When English language skills are no longer viewed as an exogenous variable emerging *ex nihilo*, but as a social feature that can co-vary with other variables, including income itself, we find that it is no longer associated with economic outcomes. (p. 262)

In other words, the level of English proficiency among a nation-state's population is an index of other factors. English language skills among a national citizenry do not necessarily advance a country's economic standing.

Other *English as development* discourses within English-language voluntourism move rapidly away from the idea of national change, however, and shift development to the individual scale. This is indeed the case with *English for entrepreneurial subjectivity* discourse.

English for Entrepreneurial Subjectivity

Contra *English for modernist development*, what I have termed *English for entrepreneurial subjectivity* discourse suggests that international development means personal development, or the creation of particular kinds of people: entrepreneurial, flexible, life-long learner-subjects who can shape-shift in response to capitalism's vagaries (see Jakubiak, 2020). In this formation, English is characterized as part of one's "bundle of skills" (Urciuoli, 2008)—a tool that one can deploy in strategic ways to enhance one's personal marketability (see also De Costa et al., 2019, as well as Schedel, this volume). *English for entrepreneurial subjectivity* discourse accords with Nelson Flores' (2013) observations about the contemporary links between pluralingualism and neoliberalism, wherein "a project of neoliberalism is to produce dynamic subjects who engage in fluid language practices that fit the needs of global capitalism" (p. 504).

Language on volunteer tourism promotional websites typically reflects some version of *English for entrepreneurial subjectivity* discourse. Marketing materials frame English language learning among individuals as a doorway or path to secure employment, financial security, and social mobility. In a statement that both illustrates *English for entrepreneurial subjectivity* discourse and casts volunteers' interventions benevolently, the NGO, Love Volunteers (2009–2023), advertises that English is becoming increasingly important among Latin American job seekers. The organization writes:

> An ability to speak English is hugely beneficial to youth, allowing them to pursue further education opportunities and to seek employment where English is an advantage, such as the tourism sector. Working as a volunteer teacher means you can help many low income children learn a tool that is vital to their future success.

Maximo Nivel (2023), another volunteer tourism sponsor, advertises its volunteer English teaching programs similarly. The organization's website states:

> People without English language skills are at an incredible disadvantage economically and in their careers. Many people dream of learning English to advance their academic and financial standing. As an international volunteer, you help make that dream become a reality.

In these excerpts, *English as development* discourse shifts away from the idea of collective, national progress through structural change. Instead,

English for entrepreneurial subjectivity discourse suggests that development occurs at the individual level by creating competitive citizen-subjects: people who can navigate economic austerity alone and shape-shift in response to market dictates. This discourse reflects a widespread view that knowing or simply studying English provides a bulwark against personal vulnerability (for related work, see Park, 2011).

English for Addressing Inequality

Links between English and equality also buttress English-language voluntourism. Alongside depictions of English language education as leading to structural transformation in a nation-state (*English for modernist development*) and shaping individual people (*English for entrepreneurial subjectivity*), a discourse that I term *English for addressing inequality* asserts that more equitable access to English language education can rectify intranational disparities among differently positioned groups. *English for addressing inequality* discourse characterizes English as a central but unevenly distributed element of individual peoples' life prospects. It suggests that if English language education were only more equitable among all people within a nation-state, uneven access to other opportunities and resources would also be reduced or eliminated.

Volunteer tourism promotional literature draws heavily upon *English for addressing inequality* discourse. NGO and commercial volunteer tourism sponsors continually frame volunteer English language teachers' service as leveling out existent disparities. In illustration, the NGO, Global Crossroad (2016a), advertises that

> the majority of Nepal's population is deprived of effective English education skills, based on a number of factors. Due to extreme poverty, children of rural families are unable to attend good schools, and fail to compete with their counterparts. For them it is either too expensive or geographically out of reach. The schools in which they can study are under-resourced and lack experienced teachers to guide them. Global Crossroad teaching English project aims to alleviate this problem by helping to make English education accessible to all.

Similarly, Maximo Nivel's (2023) website claims:

> Speaking English is a life-changing skill for people in developing countries, because the language opens both academic and professional opportunities, and provides a path out of poverty. Most middle and upper class students learn English at private language centers, which unfortunately are too expensive for a large part of the population. This creates an unfair gap between those who can afford English and those who can not [*sic*].

As we see in these examples of *English for addressing inequality* discourse, it is not unevenly structured schooling experiences that generate differential life outcomes for the disadvantaged students of a nation-state and its wealthy private school attendees. Rather, it is asymmetrical access to English language education that truncates vulnerable students' life chances.

English for addressing inequality discourse also depicts the English language as a fundamental resource for all people. Cast as on par with basic hygiene practices and emotional support, the ability to study or speak conversational English is consistently described as an "essential skill" and/or "service" in volunteer tourism promotional literature. In illustration, Global Volunteers' (2002–present) program entitled "Delivering Essential Services" lists English language teaching in a bullet-pointed list of "essential" volunteer tasks. In addition to "Teaching nutrition, hygiene, food preparation to children and adults in schools and in wor[k]shops," and "Demonstrating proper handwashing with soap and water," volunteers in Global Volunteers' "Delivering Essential Services" programs will be "Training teachers in pedagogy, English and creating supportive classroom environments" (Global Volunteers, 2002-Present).

This depiction frames English as urgently needed by the poorest and most dispossessed people. Listed alongside other urgent service tasks such as providing individuals with nutritional support and fostering personal care practices, teaching English is equated with restoring vulnerable peoples' basic health and dignity. *English for addressing inequality* discourse thus accords with Katy Highet and Alfonso Del Percio's (2021) observation that within particular social contexts, speaking English can index certain moral sensibilities such as being a clean, "hygienic" person or possessing "proper" manners or morality.

Some study participants' comments also suggested a belief that the implicit goal of English-language voluntourism is to minimize intranational wealth and opportunity gaps by teaching English to the most risk-exposed people. The discourse of *English for addressing inequality* was expressed by study participants who judged their host sites to be too well-resourced to merit volunteers' interventions. Former volunteer Samantha, for example, was frustrated because her host site in rural China had been in a fee-based English language camp rather than in a visibly poor community. Samantha said:

> I feel like the concept of volunteer teaching—you go and think, "I'm going to teach the underprivileged children, to help them." Whereas, I think the situation was with [NGO] is very much—kids pay a lot to be at the camp. And so, we weren't teaching underprivileged kids … [I]f I was to go back to China, or anywhere else to teach English, I know I wouldn't be a volunteer. Because I think the people who need it wouldn't pay, and … the whole volunteer aspect of it becomes strange. You're saying, "Okay, am I volunteering to advance the progress of China through the education of some of its citizens or what?"

Samantha's expressed incredulity that her service may have been oriented toward furthering a nation-state's broader economic position rather than helping vulnerable individuals learn English reflects *English for addressing inequality* discourse. Her stated understanding of *English as development* is that it should level out opportunity gaps within nation-states rather than among them.

Ana, another former volunteer, adopted a similar stance. She and her peers were surprised and felt cheated when they discovered that their host sites in South Africa were in middle-class, relatively affluent towns rather than visibly poor townships. To Ana, such placements were at odds with *English for addressing inequality* discourse, which had motivated her and peers to be volunteer tourists. Ana said:

> I definitely thought I would be working more in, like, a township area—like, an area that *really* needed my help ... Like, when I looked around, I just thought, "This is a really nice school." You know, what the [other] volunteer there was doing was helping kids not just learning the basics of computers, but like, really advanced programs where she worked with computers. She was like, "I didn't learn this stuff until college!" So, she was really upset that she was there tutoring these kids on something like this when she came to help kids, like, in the townships, who needed help with basic stuff like reading and writing. (emphasis in original)

Like Samantha, Ana holds a specific understanding of what English-language voluntourism is and does. In her view, volunteer interventions should follow the ideology of *English for addressing inequality* and provide people suffering from keen deprivation—those who "[need] help with basic stuff like reading and writing"—with volunteers' interventions and English lessons (see Sinervo, this volume, for similar findings).

To be sure, Samantha's and Ana's perspectives are understandable in certain respects. In the volunteer tourism gaze—one unmediated by sustained attention to structural poverty (see Crossley, 2012)—numerous aspects of vulnerability are invisible to the eye. A lack of social mobility, employment precariousness, freedom from exploitation—none of these issues correlate with visual symbols of poverty as it is constructed, often pornographically, in the West (Chouliaraki, 2013). Consequently, many volunteers deem certain communities "not poor enough" to merit outside assistance if and when those communities lack fetishized signs of material need such as houses built of corrugated metal or other re-purposed materials (see Vrasti, 2013, for similar findings).

English for addressing inequality discourse also offers the reasonable suggestion that all people, everywhere, should have equitable access to important resources. However, to cast English pronunciation or English conversational skills as equivalent to basic entitlements such as accurate public health information or medical care is problematic. For one, conversational English may be superfluous for people who work primarily in the informal economy, as many vulnerable people throughout the Global South do (Ferguson, 2013). Particularly within poor, rural communities in the Global South, English often has little social presence (Williams,

2013). In these locations, the language education most likely to impact vulnerable peoples' social futures is primary language literacy instruction. Primary language literacy skills allow vulnerable people—particularly women—to gain formal title to existent assets. By leveraging their privately held assets as collateral, women can participate in microfinance programs and substantially increase household earnings (Bruthiaux, 2002).

Moreover, English language education is always resource-intensive and can shift material and political support away from other, more critical, educational programming. In schools with underprepared teachers and few resources, English language education can have numerous, negative effects. Observing moves to expand English medium-of-instruction (MOI) policies and practices into African primary schools, for example, applied linguist Gibson Ferguson (2013) writes that

> English-language education, particularly the use of English as a primary schooling medium, has considerable costs: it reduces further the quality of what is already a weak primary education sector, it tends to advantage urban elites over the rural poor, it may constrain the development/intellectualization of local African languages and it diverts resources from under-resourced areas. Educational evidence, and efficiency considerations, point, then, to some circumscription of the educational role of English. (p. 35)

In other words, English language instruction can debilitate already-strained educational programs.

Findings from Usree Bhattacharya's (2016) ethnography of a poorly resourced village school in India affirm that English MOI policies do not reliably pull people out of poverty or eliminate entrenched educational opportunity gaps. English teachers in Bhattacharya's focal school received little preparation in English language instruction and were simultaneously in charge of large, multi-aged classes. As a result, students' English study consisted of copying rote phrases into notebooks, choral repetition of English words, and listening to translations in the local language. Students had no authentic oral exchanges in English and had few opportunities to generate meaning from English texts. Consequently, the students in Bhattacharya's focal school not only failed to learn English, but they had few opportunities to acquire basic literacy skills in *any* language. These findings lead Bhattacharya to conclude that English-medium schooling "creates further dis-citizenship for marginalized students, by offering English instruction, more or less, in name alone. In this manner, the socioeconomic disparity in India gets exacerbated through English instruction" (Bhattacharya, 2016, p. 81).

Finally, it is misleading to suggest that the mere ability to study English—in any program, with any teacher—eliminates longstanding and structured educational inequities. The quality of English language instruction within volunteer tourism is variable and usually weak (e.g., Carpenter, 2015; Henry, 2020; Jakubiak, 2016). Because little information is shared between successive volunteer tourists, for example, local students often receive repeated instruction in the English names of colors, animals, and days of the week rather than a scaffolded language program oriented toward fluency (Sinervo, 2017, and this volume). It is highly unlikely[2] that the learning outcomes of *ad hoc* English conversation practice with a rotating band of visiting volunteers could ever be equivalent to the learning outcomes of long-term, highly structured, amply resourced, bilingual and culturally relevant English language instructional programs—the programs attended almost exclusively by Global South elites (Ferguson, 2013; see also see Niño-Murcia, 2003). Yet, an aura of native speakerism (Doerr, this volume) adheres to English-language voluntourism; as we shall see, this ideology is powerful and asserts that volunteers are instructionally effective and urgently needed.

Native English Immersion as Pedagogy

In all the versions of *English as development* that circulate within English-language voluntourism, English is described as a technical tool that must

[2] Attempts, such as they are, to determine English-language voluntourism's instructional efficacy are almost non-existent. Studies of volunteer tourism more broadly often mention that the English language instruction offered by visiting volunteers is unassessed (cf. Carpenter, 2015) and boring for students (Everingham, 2018; Sinervo, this volume). Participants in the volunteer programs under examination here report receiving little on-site instructional support, are often given few curricular resources, and express feelings of being overwhelmed by the twin demands of lesson planning and creating a classroom community (see Jakubiak, 2016).

conform to exogenous norms. In what I term *native English immersion as pedagogy* discourse, such linguistic instrumentalism (Wee, 2008) is supplemented by continual references to an idealized "native speaker": someone who speaks a prestige-variety of English and is even, at times, explicitly "Western." *Native English immersion as pedagogy* discourse suggests that visiting volunteers are urgently needed throughout the Global South for two main reasons: (1) to provide an immersive language learning experience for disadvantaged students; and (2) to model prescriptive language use and innovative instructional techniques to local language teachers.

Native English immersion as pedagogy discourse is featured prominently on English-language voluntourism sponsors' websites. The NGO, Globe Aware (n.d.), for example, frames having little contact with native English language speakers as a kind of community impoverishment. Describing a volunteer's role in its Cambodia program, the organization states:

> Volunteers will participate in a variety of projects … working with children in a variety of interactive opportunities at a village school in a rural Cambodian village who otherwise have very little contact with native English speakers.

Similarly, the NGO, Global Crossroad (2016b), describes its short-term program in Cambodia to prospective volunteers as follows:

> [Y]ou are providing an incredibly valuable resource to the schools, and through this great value to the kids themselves as an opportunity to learn English from someone who's native or fluent will greatly improve their chances at a thriving future.

Across these promotional materials, having "little contact with native English speakers" or not being able to "learn from someone who's native" are framed as forms of educational deprivation. This rhetoric suggests that without a visiting, native English language speaker, host community members cannot learn English and will suffer economically as a result.

A distinct feature of *native English immersion as pedagogy* discourse is its congruence with *native speaker saviorism* (Jenks & Lee, 2020).

English-language voluntourism discourse continually offers that regardless of one's educational background or prior work experience, if one is a (Western) native English language speaker, one can provide high-quality English instruction in any context and further boost learners' confidence. Here, the NGO, The Friends for Asia Foundation (n.d.), uses *native English immersion as pedagogy* discourse to explain how visiting, short-term English language teachers can foment social change through their native speaker status:

> Many Thai students grow up thinking that their English ability is next to nil. But sometimes, a few confidence-boosting classes with a native speaker—where they're understanding more than they expected to and communicating right back—is all it takes to change their minds. As a volunteer English teacher, you become a catalyst for student development.

Whether the providers of self-esteem, the "correct" accent, or innovative teaching techniques, English-language voluntourists are continually cast in *native English immersion as pedagogy* discourse as model language users and instructional experts. The discourse draws upon *native speaker saviorism* by positioning prospective volunteer tourists as nearly obliged to share their English native speakerness with marginalized people around the world.

Study participants also spoke in ways that reflected *native English immersion as pedagogy* discourse. Some volunteers constructed a hierarchy of global English(es) in which their own, prestige varieties were on top. Shannon, for example, a former Namibia volunteer, described her host community's language-in-use practices as follows:

> In a lot of the classes, one thing I noticed is that a lot of the teachers were teaching in Damara, which is the local language, and when they taught in English, they didn't teach in proper English. They taught in the like, local, like dialect English, if that makes sense. Like, they were using like the colloquialisms and the improper grammar that, like, is Namibian English.

Shannon's understanding is that Namibian English is an illegitimate variety of English (see Reagan, 2016, on *language legitimacy*). Although volunteer tourism sponsors continually stress that these programs offer volunteers a rich opportunity to learn about other cultures and even become global citizens, Shannon's comments suggest she remains unfamiliar with numerous, arguably global, sociolinguistic phenomena such as language creolization, linguistic hybridity, and World Englishes, to name a few (see Pennycook, 2020).

Native English immersion as pedagogy discourse also appeared in volunteers' judgments of host teachers' instructional practices. Morgan, an in-service volunteer, for example, characterized her host school's English program director as pedagogically uninspiring and inattentive. During our interview at the NGO's mid-service conference in Costa Rica, Morgan contrasted her own service role with the efforts of the local program director:

> [My] going and teaching English is really helping them [local students]. Like, I'm a native speaker; I can come there and … I can give that extra effort these next few, few months—actually, weeks now—that the director couldn't give them. So, it's definitely like, a me giving to them what their other—their teachers really can't or just don't really do, like, really engaging with them during class. Because what I've experienced is like my director would just, like, sit down and the kids would do a worksheet.

Here, Morgan frames her language instruction as superior to that of her Costa Rican host. She sees herself as an active, inspiring educator in contrast to business-as-usual. Although the teacher whom Morgan describes is also the English program director and presumably responsible for a myriad of administrative duties that Morgan is not, Morgan casts him as unable to engage with students and therefore instructionally ineffective.

Audrey, a former Ecuador volunteer, employed *native English immersion as pedagogy* discourse similarly. Like Morgan, Audrey had no prior experience as a language educator or K-12 classroom teacher. However, she positioned herself as knowing more about language pedagogies than her Ecuadorian hosts. When asked to describe her understanding of English-language voluntourism's broader purposes and goals, Audrey explained that visiting volunteers model active and innovative instructional techniques. She said:

> I think [the NGO was] trying to make the education system better, because it was just terrible. And, of course they can't do it top down—they can't just tell the government to just change, I mean, their own—but if they send a volunteer to send and show, like, lead by example—be like, "Look how much the kids learn when they play the game instead of, like, yelling out sentences," you know? Then teachers would kind of like, learn ... And as soon as [local teachers] saw the things that were good, mostly—I mean, some of them were like, "Whoa, whoa, I hate games" and didn't do it ... but like, just like, meaning like the pointers, like, "Hey, here's another way to think about it, you know, let's play like, you know, 'Capture the Flag' in English. Like, wow, they all know how to say 'flag' now, or, like, 'You're it.'" I think just leading by example.

Consistent with *native English immersion as pedagogy* discourse, Audrey casts herself as bringing energy and enthusiasm to staid local English language classrooms. In her perspective, playing games such as tag or "Capture the Flag" are examples of progressive language teaching. This view aligns with research demonstrating that Western-centric, communicative language teaching methods are often fetishized in English language teaching (ELT) contexts (e.g., Canagarajah, 2002; Haque & Morgan, 2009). Additionally, both Morgan's and Audrey's portrayals of themselves as bringing new energy to tired host classrooms reflect what Jacob Henry (2020) refers to as the "maverick" cinematic underprepared teacher narrative, which "devalues already-existing practices within the host school. The outsider comes in to shake things up and bring new perspectives, but assumes that the [local] school operates only with deficiency" (p. 4).

Discrepant Data

While most of my study participants took up *native English immersion as pedagogy* discourse's claims, a few specifically rejected them. In retrospective interviews, some former volunteers tied their own instructional ineffectiveness to English immersion pedagogies. Chris, for example, a former

Ecuador volunteer, explained that he had followed the NGO's prescribed way of teaching—in English-only—and later thought differently about it. The following exchange between me and Chris illustrates his shifting feelings about "immersion":

Cori: What was your understanding of how you should teach English—that you got from the organization there or during orientation?

Chris: The main thing that was, I think, kind of repeated over and over was, like, complete immersion. Like, there should be no Spanish in the classroom. Just because, like, it's a total immersion experience, then they'll learn more that way. And I think that I didn't speak Spanish in the classroom except when I had to, like, yell at people,[3] like, after class and they didn't understand English. But, I don't—I don't know if that's actually the best method.

Cori: Why's that?

Chris: I feel like I could've covered a lot more and explained things, like, with more clarity if I could've—if I would've used some Spanish. Like, we spent so—we spent way too long, like, going over cognates and what is a cognate. Whereas, like, if I would've done it in Spanish, it would've been just, like, two minutes.

Here, Chris rejects the dominant precepts of *native English immersion as pedagogy* discourse. He laments that simple explanations took too long and regrets not using more Spanish.

Josh, another former volunteer, also expressed skepticism toward *native English immersion as pedagogy* discourse. In a retrospective interview, he explained that he has kept in touch with his former Costa Rican host family, who expressed disappointment that his volunteer successor didn't speak any Spanish. This successor's lack of Spanish language skills detracted from his overall influence, in Josh's view. Josh said:

[3] The comment that Chris makes here about using Spanish to "yell at people" offers insight on how English-language voluntourism operates in host classrooms. Because volunteers receive little instruction on how to build positive classroom communities, they are often stressed by their perceived lack of "control" and employ top-down, teacher-centered, management strategies (see also Henry, 2020).

> It was sort of strange, because, you know, they [my former host family] ... they were saying that the volunteer after me didn't speak any Spanish, so it was difficult to communicate with [him] because my community was—there were almost no English speakers. My [host] family didn't speak any English. So, I guess they said it was difficult in that respect. And he [the volunteer] would be reading all the time, so he wasn't out talking with people, which I guess wasn't as much his fault as his lack of, you know, ability to speak Spanish ... [My host family] had a lot better vision of me because I could actually communicate with them to a limited extent and explain what I was thinking, explain things about my culture. Whereas if someone just went there and spoke total English—as [the NGO] almost had instructed us to do—I don't know if that cultural exchange would have happened as much. I mean, obviously, the classroom is different from the home, but at least in the—even in the classroom, I used a lot more Spanish than I thought I would in order to sort of get them to respect me, almost. And then, you know, often when you're teaching, you need to explain things in a story form or do some—do something like that or start making fun of people or make jokes, so I thought that my Spanish helped me make a better impact.

Josh's reflections point to the limitations of *native English immersion as pedagogy* discourse. Not knowing or using the local, or community, language can prohibit volunteers from connecting with students and forming meaningful relationships with host communities. This position is aligned with the work of Phoebe Everingham (2018), who argues that translingual exchanges between volunteer tourists and community hosts can foster increased interpersonal connections (see also Sinervo, this volume).

Peg, a former volunteer in Namibia, also voiced a stark critique of *native English immersion as pedagogy* discourse. A liberal arts major at a prestigious Ivy League college, Peg reflected on her time as a volunteer English teacher in Namibia with a great sense of humility. She said:

> I struggled with the fact that I didn't speak any Oshikwanyama. And like, by the end of it, I was like, "This is terribly disrespectful for me to, like, come to their, you know, their town and just expect that I could get along there without speaking any Oshikwanyama." And like just expecting that

> everyone would speak English all the time, which obviously, they didn't. They probably spoke it, like, ten to twenty percent of the time—when I was even in the room.

Here, Peg expresses regret at having traveled to Namibia without possessing any communicative competence in the local language, Oshikwanyama. Her sense that it was "terribly disrespectful" to be ignorant of the local language again reflects Everyingham's (2018) findings that host community members in volunteer tourism contexts appreciate and value volunteers' attempts to learn local languages (see also Sinervo, this volume). Yet, the discourse of *native English immersion as pedagogy* exalts this very form of ignorance. It denies the role of host community languages as resources for language learning, and it ignores how translingual exchanges can help volunteers to connect meaningfully to host community members.

Discussion

I began this chapter by asking why and how English-language voluntourism is so widely understood to be a useful and necessary intervention. In tracing the various discourses that adhere to English-language voluntourism, my aim has been to highlight how volunteer tourism sponsors as well as in-service and former volunteers all draw upon particular discursive frameworks to justify the practice. Although the four *English as development* discourses described here often overlap, taken together, they generate a warrant for English-language voluntourism that is nearly unassailable. If (1) nation-states need more English in order to be competitive economically; (2) individual people need English language skills to find and maintain secure employment; (3) students in under-resourced schools need English language lessons to keep up with their better-schooled peers; and (4) existent English language instruction in the Global South could benefit from native English language speakers' presence and innate instructional knowledge, what could be *wrong* with English-language voluntourism? And what does English-language voluntourism's popularity tell us about the present moment?

Like other studies of segmented volunteer tourism, my analysis of English-language voluntourism sheds light on contemporary modes of neoliberal governance and the structural inequality it creates. By drawing upon discourses that cast English as needed at multiple scales (the national *and* the individual), English-language voluntourism bolsters phenomena such as *English fever* (see Choi, this volume, as well as Jang, this volume), which naturalizes English language education as a response to shrinking labor markets and increased economic precarity. This precarity has elevated not only English language competency but also the simple *study* of English as a new marker of distinction among job seekers. A person's relations to the English language—whether they know it, have studied it, or can use it in various contexts—now serve as an index of their self-regulatory spirit (see Gao, 2019) or even basic "personhood" (see Highet & Del Percio, 2021). Many job seekers now "invest" in English with no guarantee of a return (see Kubota, 2011), and English-language voluntourism operates as a site for this speculation.

English-language voluntourism also "works" because it harnesses the biopolitical imperative that Western young people—the volunteers themselves—become active, independent, and intrepid global citizens through international experiences abroad (see McDaid & Sunyol, this volume, as well as Doerr, this volume). Both *English for addressing inequality* and *native immersion as pedagogy* discourse frame inexperienced, native English language speakers as simultaneously meeting a global need while broadening their own horizons. Narrow ideas of language pedagogy, such as "immersion," provide ideological cover for the presence of Inner Circle, often "native," English language speakers in classrooms worldwide, where host community members are cast as suffering for their dual lack of English as well as native English language speaker role models. This warrant for *native speaker saviorism*, as Christopher J. Jenks & Jerry Won Lee (2020) point out, "reflects the long-standing assumption that the White community can 'save' peoples of color by teaching them English" (p. 186), and English-language voluntourism is deeply rooted in this saviorist imperative.

Former volunteers' critical reflections, however, suggest a space of possibility. My data reveal that not all volunteers take up English-language voluntourism's dominant discourses uncritically, and some former volunteers in my study specifically rejected *native English immersion as pedagogy* discourse. Chris, Josh, and Peg all saw the limitations of English-only ideologies, and Josh, and in particular, saw Spanish language use as a way to connect to his host community. Relatedly, Peg was ashamed of her lack of basic communicative competency in Oshikwanyama. The self-consciousness and self-awareness among volunteers on display here suggest that developing *critical language awareness* (Alim, 2005) can be a powerful and more linguistically just outcome of participating in English-language voluntourism. Languages and their speakers can come into contact in surprising and revelatory ways in voluntourism encounters, which show that the outcomes of English-language voluntourism are not predetermined.

Conclusion

While numerous writers continue to debate whether volunteer tourism "does more harm than good" or offer suggestions on how these programs might be "fixed," studies of language-motived voluntourism, as we see in this volume, are moving in new directions. The analysis that I have presented here reveals how particular discursive frames come together to present English-language voluntourism as helpful; to justify the mobilities and classroom interventions of native English language speakers; and to buttress broader, neoliberal arguments about English's importance in the world (see O'Regan, 2021, for more on global English language spread and political economy). My work is not meant to halt or improve English-language voluntourism programs. Rather, it attempts to underscore how different discourses and ideologies come together to imbue particular English varieties, speakers, and global mobilities with high (and exchangeable) value in English-language voluntourism. Whether things will always be this way remains to be seen. In the face of rising movements for linguistic justice, for example, one wonders whether these programs' popularity will hold.

Finally, this chapter has also highlighted the critical need for more studies of segmented volunteer tourism. Unlike volunteer tourism in animal sanctuaries or refugee resettlement contexts, English-language voluntourism reveals specific features of the present such as the role of language in the White savior industrial complex (see also Jenks & Lee, 2020). In that sense, segmented volunteer tourism studies allow us to better trace how even seemingly benign interventions can uphold broader structural inequalities. More careful and targeted analyses of specific voluntourism forms—be these language-motivated, refugee resettlement-based, or related to animal sanctuaries—may expand our theoretical toolboxes at a time when we very much need them.

Acknowledgments I offer my sincere gratitude to Larissa Schedel and Greta Schmidt for their constructive feedback on multiple drafts of this chapter.

Research Funding This research was supported by both the Graduate School and the Latin American and Caribbean Studies Institute at the University of Georgia as well as the Committee for the Support of Faculty Scholarship at Grinnell College.

References

Alim, H. S. (2005). Critical language awareness in the United States: Revisiting issues and revising pedagogies in a resegregated society. *Educational Researcher, 34*(7), 24–31. https://doi.org/10.3102/0013189X034007024

Appleby, R., Copley, K., Sithirajvongsa, S., & Pennycook, A. (2002). Language in development constrained: Three contexts. *TESOL Quarterly, 36*(3), 323–346. https://doi.org/10.2307/3588416

Arcand, J. L., & Grin, F. (2013). Language in economic development: Is English special and is linguistic fragmentation bad? In E. J. Erling & P. Seargeant (Eds.), *English and development: Policy, pedagogy and globalization* (pp. 243–266). Multilingual Matters.

Bernstein, J. D., & Woosnam, K. M. (2019). Same same but different: Distinguishing what it means to teach English as a foreign language within the context of volunteer tourism. *Tourism Management, 72*, 427–436. https://doi.org/10.1016/j.tourman.2019.01.010

Bhattacharya, U. (2016). The politics of participation: Dis-citizenship through English teaching in a suburban Indian village school. *English as an International Language Journal, 11*(1), 71–85.

Bruthiaux, P. (2002). Hold your courses: Language education, language choice, and economic development. *TESOL Quarterly, 36*(3), 275–296. https://doi.org/10.2307/3588414

Bunce, P. (2016). Voluntary overseas English language teaching: A myopic, altruistic hydra. In P. Bunce, R. Phillipson, V. Rapatahana, & R. Tupas (Eds.), *Why English? Confronting the hydra* (pp. 106–117). Multilingual Matters.

Canagarajah, A. S. (2002). Globalization, methods, and practice in periphery classrooms. In D. Block & D. Cameron (Eds.), *Globalization and language teaching* (pp. 134–150). Routledge.

Carpenter, N. (2015). Childhood studies and orphanage tourism in Cambodia. *Annals of Tourism Research, 55*, 15–27. https://doi.org/10.1016/j.annals.2015.08.010

Charmaz, K. (2006). *Constructing grounded theory: A practical guide through qualitative analysis*. Sage.

Chouliaraki, L. (2013). *The ironic spectator: Solidarity in the age of post-humanitarianism*. John Wiley & Sons.

Cole, T. (2016). *Known and strange things: Essays*. Random House.

Collard, R. C. (2020). *Animal traffic: Lively capital in the global exotic pet trade*. Duke University Press.

Crossley, É. (2012). Poor but happy: Volunteer tourists' encounters with poverty. *Tourism Geographies, 14*(2), 235–253. https://doi.org/10.1080/14616688.2011.611165

De Costa, P. I., Park, J. S.-Y., & Wee, L. (2019). Linguistic entrepreneurship as affective regime: Organizations, audit culture, and second/foreign language education policy. *Language Policy, 18*(3), 387–406. https://doi.org/10.1007/s10993-018-9492-4

Erling, E. J., & Seargeant, P. (Eds.). (2013). *English and development: Policy, pedagogy and globalization*. Multilingual Matters.

Everingham, P. (2018). Speaking Spanglish: Embodying linguistic (b)orderlands in volunteer tourism. *Emotion, Space and Society, 27*, 68–74. https://doi.org/10.1016/j.emospa.2018.04.001

Ferguson, G. (2013). English, development, and education: Charting the tensions. In E. J. Erling & P. Seargeant (Eds.), *English and development: Policy, pedagogy and globalization* (pp. 21–44). Multilingual Matters.

Flores, N. (2013). The unexamined relationship between neoliberalism and plurilingualism: A cautionary tale. *TESOL Quarterly, 47*(3), 500–520. https://doi.org/10.1002/tesq.114

Gao, S. (2019). *Aspiring to be global: Language and social change in a tourism village in China*. Multilingual Matters.
Global Crossroad. (2016a). *Volunteer in Nepal!* Retrieved May 1, 2023, from https://www.globalcrossroad.com/nepal/volunteer-in-nepal
Global Crossroad. (2016b). *Volunteer Teaching in Cambodia!* Retrieved May 1, 2023, from https://www.globalcrossroad.com/cambodia/volunteer-teaching-in-cambodia
Global Volunteers. (2002–present). Delivering essential services with global Volunteers. Retrieved September 5, 2022, from https://globalvolunteers.org/delivering-essential-services/
Globe Aware. (n.d.). *Volunteer vacations in Cambodia*. Retrieved May 1, 2023, from https://globeaware.org/destinations/asia/cambodia
Grout, P. (2009). *The 100 best volunteer vacations to enrich your life*. National Geographic Books.
Haque, E., & Morgan, B. (2009). Un/marked pedagogies: A dialogue on race in EFL and ESL settings. In R. Kubota & A. Lin (Eds.), *Race, culture, and identities in second language education: Exploring critically engaged practice* (pp. 271–285). Routledge.
Henry, J. (2020). The cinematic pedagogies of underprepared teachers. *Teaching and Teacher Education, 89*, 1–9. https://doi.org/10.1016/j.tate.2019.102990
Highet, K., & Del Percio, A. (2021). When linguistic capital isn't enough: Personality development and English speakerhood as capital in India. In J. E. Petrovic & B. Yazan (Eds.), *The commodification of language: Conceptual concerns and empirical manifestations* (pp. 127–143). Routledge.
Jakubiak, C. (2016). Ambiguous aims: English-language voluntourism as development. *Journal of Language, Identity & Education, 15*(4), 245–258. https://doi.org/10.1080/15348458.2016.1195270
Jakubiak, C. (2020). "English is out there—You have to get with the program": Linguistic instrumentalism, global citizenship education, and English-language voluntourism. *Anthropology & Education Quarterly, 51*(2), 212–232. https://doi.org/10.1111/aeq.12332
Jakubiak, C. (2021). The possibility of critical language awareness through volunteer English teaching abroad. In K. Swalwell & D. Spikes (Eds.), *Anti-oppressive education in "elite" schools: Promising practices and cautionary tales from the field* (pp. 109–121). Teachers College Press.
Jenks, C. J., & Lee, J. W. (2020). Native speaker saviorism: A racialized teaching ideology. *Critical Inquiry in Language Studies, 17*(3), 186–205. https://doi.org/10.1080/15427587.2019.1664904

Kamwangamalu, N. (2008). Language policy, vernacular education and language economics in postcolonial Africa. In R. Rubdy & P. K. W. Tan (Eds.), *Language as commodity: Global structures, local marketplaces* (pp. 171–186). Continuum.

Kapoor, I. (2004). Hyper-self-reflexive development? Spivak on representing the third world 'other'. *Third World Quarterly, 25*(4), 627–647. https://doi.org/10.1080/0143659041000167889‌8

Knight, S. (2023). Volunteer Latin America: Honeyteering – Volunteering for honeymooners. Retrieved October 5, 2022, from https://www.volunteerlatinamerica.com/blog/posts/honeyteering

Knott, A. (2018). Guests on the Aegean: Interactions between migrants and volunteers at Europe's southern border. *Mobilities, 13*(3), 349–366. https://doi.org/10.1080/17450101.2017.1368896

Kubota, R. (2011). Learning a foreign language as leisure and consumption: Enjoyment, desire, and the business of *eikaiwa*. *International Journal of Bilingual Education and Bilingualism, 14*(4), 473–488. https://doi.org/10.1080/13670050.2011.573069

Love Volunteers. (2009–2023). *Teach English overseas*. Retrieved April 30, 2023, from https://www.lovevolunteers.org/destinations/volunteer-guatemala/education-support-antigua-and-surrounding-areas-mayan-highlands

Marcus, G. E. (1995). Ethnography in/of the world system: The emergence of multi-sited ethnography. *Annual Review of Anthropology, 24*(1), 95–117. https://doi.org/10.1146/annurev.an.24.100195.000523

Maximo Nivel. (2023). Volunteer teaching English: Education volunteer projects abroad. Retrieved April 30, 2003, from https://maximonivel.com/volunteers/teaching-english/

McMillion, B., Cutchins, D., & Geissinger, A. (2012). *Volunteer vacations: Short-term adventures that will benefit you and others*. Chicago Review Press.

New Hope Volunteers. (2015). *Volunteering abroad!* Retrieved April 15, 2022, from http://newhopevolunteers.org

Niño-Murcia, M. (2003). "English is like the dollar": Hard currency ideology and the status of English in Peru. *World Englishes, 22*(2), 121–141. https://doi.org/10.1111/1467-971X.00283

O'Regan, J. P. (2021). *Global English and political economy*. Routledge.

Park, J. S.-Y. (2011). The promise of English: Linguistic capital and the neoliberal worker in the south Korean job market. *International Journal of Bilingual Education and Bilingualism, 14*, 443–455. https://doi.org/10.1080/13670050.2011.573067

Parreñas, R. J. S. (2012). Producing affect: Transnational volunteerism in a Malaysian orangutan rehabilitation center. *American Ethnologist, 39*(4), 673–687. https://doi.org/10.1111/j.1548-1425.2012.01387.x

Peet, R., & Watts, M. (1993). Introduction: Development theory and environment in an age of market triumphalism. *Economic Geography, 69*(3), 227–253. https://doi.org/10.2307/143449

Pennycook, A. (2020). Translingual entanglements of English. *World Englishes, 39*(2), 222–235. https://doi.org/10.1111/weng.12456

Phillipson, R. (1992). *Linguistic imperialism*. Oxford University Press.

Quesada, S. (2014). Development and its social, economic, and educational consequences: The case of the Zimapán hydroelectric project. In P. R. Portes, S. Salas, P. Baquedano-López, & P. J. Mellom (Eds.), *U.S. Latinos and education policy: Research-based directions for change* (pp. 115–125). Routledge.

Reagan, T. (2016). The conceptualization of language legitimacy. *Critical Inquiry in Language Studies, 13*(1), 1–19. https://doi.org/10.1080/15427587.2015.1116950

Schedel, L. S. (2021). The price of immersion: Language learners as a cheap workforce in Malta's voluntourism industry. *Multilingua, 41*(2), 181–200. https://doi.org/10.1515/multi-2020-0178

Schwarz, K. C., & Richey, L. A. (2019). Humanitarian humor, digilantism, and the dilemmas of representing volunteer tourism on social media. *New Media & Society, 21*(9), 1928–1946. https://doi.org/10.1177/1461444819834509

Sinervo, A. (2017). Interdependent industries and ethical dilemmas: NGOs and volunteer tourism in Cusco, Peru. In A. Lashaw, C. Vannier, & S. Sampson (Eds.), *Cultures of doing good: Anthropologists and NGOs* (pp. 142–162). The University of Alabama Press.

Stainton, H. (2016). A segmented volunteer tourism industry. *Annals of Tourism Research, 61*(C), 256–258. https://doi.org/10.1016/j.annals.2016.09.011

The Friends for Asia Foundation. (n.d.). *English education volunteer - Bangkok and Chiang Mai*. Retrieved August 27, 2003 from https://www.volunteer-thailand.org/english-education-chiang-mai/

Thomas-Maude, J., McLennan, S., & Walters, V. (2021). Cultural exchange during English-language voluntourism (EVT) in Lima, Peru: A postcolonial analysis. *Journal of Language, Identity & Education*, 1–16. Advance online publication, https://doi.org/10.1080/15348458.2021.1991802

Tobin, J., & Hayashi, A. (2017). Return interviews and long engagements with ethnographic informants. *Anthropology and Education Quarterly, 48*(3), 318–327.

Urciuoli, B. (2008). Skills and selves in the new workplace. *American Ethnologist, 35*(2), 211–228.
Volunteer Latin America. (2022). *Find free and low-cost volunteer opportunities throughout Central and South America.* Retrieved September 10, 2022 from https://www.volunteerlatinamerica.com
Vrasti, W. (2013). *Volunteer tourism in the global south: Giving back in neoliberal times.* Routledge.
Watkins, A. (2013). Expert cruise advice: Volunteering on a cruise. *Cruise International.* https://www.cruiseandtravel.co.uk/volunteering-on-a-cruise/
Wearing, S. (2001). *Volunteer tourism: Experiences that make a difference.* CABI.
Wee, L. (2008). Linguistic instrumentalism in Singapore. In P. Tan & R. Rubdy (Eds.), *Language as commodity: Global structures, local marketplaces* (pp. 31–43). Continuum.
Williams, E. (2013). Political perspectives on language policies and development in Africa. In E. J. Erling & P. Seargeant (Eds.), *English and development: Policy, pedagogy and globalization* (pp. 68–87). Multilingual Matters.

3

Becoming "TEFL Certified": Professionalization, Certification, and Commodification in Teaching English as a Foreign Language Volunteer Tourism

Joshua D. Bernstein

Introduction

Despite the growth of the volunteer tourism industry, concern remains regarding unqualified participants and their ability to "help" disadvantaged communities. In English language teaching (ELT), one's decision to work or volunteer is often based on their background, where those with experience seek employment and those without, volunteer to acquire experience (Bernstein & Woosnam, 2019). This results in under-resourced schools, institutions in need of the most qualified and committed teachers, often being the primary destination of short-term untrained volunteer tourists.

J. D. Bernstein (✉)
Thammasat University, Bangkok, Thailand
e-mail: joshua.b@litu.tu.ac.th

L. S. Schedel, C. Jakubiak (eds.), *Voluntourism and Language Learning/Teaching*,
Palgrave Advances in Language and Linguistics,
https://doi.org/10.1007/978-3-031-40813-7_3

Faced with increased public and academic pressure to "professionalize" teaching in volunteer tourism, teaching English as a foreign language (TEFL) certificate[1] courses have begun to gain popularity. Defined as "a course of study delivered in English which prepares candidates to teach EFL and/or ESL to adult, adolescent, or young learners" (Staehr Fenner & Snyder, 2014, p. 4), TEFL certification has traditionally functioned as a bridge into the ELT profession. While entry-level ELT jobs have accepted this form of training with limited discretion (Tanner, 2003), more recently, online TEFL certificates are being advertised to potential volunteer tourists as a way improve their pedagogical knowledge before embarking on a service trip (e.g., Go Overseas, 2020; TEFL Institute of Ireland, 2022). International Volunteer HQ (n.d.), for example, states:

> We encourage volunteers on IVHQ's [International Volunteer HQ] Teaching projects to come prepared by completing some relevant training, such as a Teaching English as a Foreign Language (TEFL) course. Although formal teaching qualifications are not required on these projects as volunteers are supported by local staff, taking a TEFL course *enables you to gain more from the experience* [author's emphasis].

Sending organizations suggesting volunteers equip themselves with the knowledge and skills to successfully benefit a host community is a new development in an industry that has historically not required any teaching credentials (Jakubiak, 2020).

Conscious and thoughtful pre-service preparation covering declarative (knowing what) and procedural (knowing how) knowledge of language, methodology, and interculturality (Stanley & Murray, 2013) combined with critical, social justice-oriented teacher education which encourages reflection and discussion (Anya, 2018; Nygreen et al., 2015) could appeal to those looking to enhance the effectiveness of TEFL volunteer tourism. At the same time, however, incongruence within the ELT industry provides profit-driven companies the "credence to trumpet flamboyant (yet empty) language and subsequently capitalize on this ambiguity" (Bernstein, 2021, p. 4). Volunteer tourists caught up in the

[1] Certificates go by a variety of different acronyms (e.g., TESL, TESOL).

commodification of travel and pursuit of unique life experiences (e.g., gap year-type breaks) may similarly seek "to gain more" by employing their economic capital (i.e., buying a TEFL certificate course) to acquire social and cultural capital (e.g., soft skills, international experience), thus enhancing themselves (Brown et al., 2003; Heath, 2007) but not necessarily benefiting those they teach.

Concerned with the consumption of a packaged TEFL certificate and the framing of education as a product for sale, this chapter makes a critical contribution to the volunteer tourism/language education nexus by building on the discussion of commodification in sustainable/alternative forms of tourism and critique of volunteer tourism as being "unprofessional" (e.g., Coren & Gray, 2012; Gray & Campbell, 2007; Wearing, 2001). While commercial sponsors encouraging ELT preparation is a promising step forward, whether these courses offer effective, culturally sustaining language teacher education has not been explored.

Literature Review

Teaching English is a common type of volunteer tourism (Bernstein & Woosnam, 2019; Jakubiak, 2016; McMillon et al., 2012) and is "a practice in which Global North, often young and inexperienced, volunteers teach English in the Global South on a short-term basis as a form of alternative travel" (Jakubiak, 2020, p. 212). The niche lays at the crossroads of tourism, volunteering, and the ELT profession (Bernstein & Woosnam, 2019; Stainton, 2018a, 2018b). While research on TEFL volunteer tourism is only in its nascent stage, scholars have begun to unravel its complexity. Joshua Bernstein and Kyle Woosnam (2019), for example, found TEFL volunteer tourism and TEFL job advertisements differed among their stated qualifications, benefits, and description. Additionally, Hayley Stainton (2018a) introduced the term TEFL tourist[2] to describe "a person who travels outside of their usual environment to teach English

[2] In essence this is a repackaging of the pejorative "backpacker teacher," typically characterized as an unqualified, White, "native English speaker" from an Inner Circle country who extends their trip in an Expanding Circle country through leveraging their privileged status to gain employment as an English teacher (see Copland et al., 2016; Thornbury, 2001).

as a foreign language, whose role shifts between tourist, educator and educatee at various points in their trip" (p. 2).

TEFL volunteer tourism has developed in response to a shortage of teachers (Stainton, 2018b) and the belief that "native speakers" possess the ability to intuitively teach English (Robert Phillipson, 1992). Research also suggests that neoliberalizing economies contribute to the demand for English worldwide, particularly Inner Circle varieties which command the highest status in the global ELT marketplace (see Choi, this volume; Park, 2021). Other forces that have contributed to the rise of TEFL volunteer tourism include the continued salience of Whiteness in ELT (Jenks & Lee, 2020) and increasing biopolitical mandates that link English to economic stability, personal growth, and self-actualization (De Costa et al., 2016).

"Doing Well" by (Purportedly) "Doing Good"

TEFL volunteer tourism appeals to the idealism, ambition, and self-interest of socially committed individuals who want to "do well" while "doing good" (Labaree, 2010). International volunteers "do well" by participating in a socially valued experience that helps craft a desirable "personality package" (Heath, 2007) based on the development of intrapersonal and interpersonal skills (Raymond, 2008; Wearing, 2001). These intangible benefits are presumed a byproduct of the interaction and exchange volunteers can have with host community members (Raymond & Hall, 2008) and are promoted by sponsoring organizations (Lyons et al., 2012).

While the question of whether volunteer tourists (and their sending organizations) are "doing well" can be answered by observing the industry's growth, whether (often) young, inexperienced Westerners are "doing good" has been central to public and academic debate (e.g., Palacios, 2010; Sin, 2010; Tomazos, 2021). Following foundational research on volunteer tourism which mainly advocated its benefits, academic investigation began to take a more cautionary approach (Wearing et al., 2017). The extant literature outlined a number of voluntourism's potential negative impacts (see Guttentag, 2009).

Concern for the lack of skills generally needed to participate in volunteer tourism has been central to this critical body of inquiry (e.g., Brown & Hall, 2008; Raymond & Hall, 2008) and subsequent research seeking to foster more equitable exchange between volunteer and host (Wearing & McGehee, 2013). As observed by Sally Brown and Alastair Morrison (2003), "the only skill that is required by many of these organizations is the desire to help others" (p. 77). This lack of prerequisites has helped attract volunteers and contributed to the industry's growth but also disproportionately focuses on the desires of sending organizations and volunteers, resulting in questionable host community benefits (Guttentag, 2009; Palacios, 2010; Raymond, 2008).

Dominant critiques of volunteer tourism also call for the acknowledgment of power, structural inequality, and privilege in the colonial-esque encounters where volunteers are often inappropriately framed as "experts" (Raymond & Hall, 2008; Simpson, 2004; Sin, 2010). According to Frances Brown and Derek Hall (2008):

> The use of volunteers, who often have little knowledge or experience of the work they are undertaking (an attraction for the volunteers), also calls into question their effectiveness and raises the spectre of neo-colonialism in the tacit assumption that even ignorant Westerners can improve the lot of people in the South. (p. 845)

While volunteer tourism has the ability to produce mutually positive economic/social outcomes, issues of Western privilege and power are often not addressed. Instead, "simplistic binaries of 'us and them'" are perpetuated (Simpson, 2004, p. 690), thereby reinforcing power inequalities and doing very little with regard to structural change (Lyons et al., 2012; Sin, 2010; Wearing, 2001).

Online TEFL certificates have emerged as a way to capitalize on the growing TEFL volunteer tourism market, respond to criticism regarding volunteer effectiveness, and improve the planning/management of gap year organizations (Bernstein & Woosnam, 2019; Frilund, 2018; Guttentag, 2009; Lyons et al., 2012; Raymond, 2008; Simpson, 2004; Sinervo, 2017; Tomazos, 2021). As TEFL certificates are increasingly being marketed to volunteer tourists, it is important to consider their merit and if these programs contribute to the broader commodification of language education (Heller, 2010).

TEFL Certification

Short-term TEFL certificates are the quickest and easiest way to enter the ELT profession (Kanowski, 2004). While their online manifestation represents a fairly recent phenomenon, various iterations of in-person pre-service courses have existed for more than half a century (Thomson, 2004). As Diane Staehr Fenner and Sidney Snyder (2014) summarize, "programs are typically 120 to 180 interactive hours in length and include: (1) instruction in language and culture, (2) instruction in TESOL theory and methodology, and (3) classroom observation and supervised practice teaching with English language learners" (p. 4).

One of the first formalized short-term TEFL courses was taught in 1962 by John and Brita Haycraft, the founders of International House London[3] (International House London, n.d.). Whereas in-service/post-experience courses were available at the diploma or master's level, initial training at the time was done by private sector institutions (Howatt, 1984). Underwhelmed by the existing theory-based approaches to teacher education, the Haycrafts developed a practical two-week pre-service course at their language school (Haycraft, 1988). Through blending theory with practice, this pioneering course helped raise ELT standards and evolved into what is today the Cambridge Certificate in Teaching English to Speakers of Other Languages (CELTA), which is "the most widely recognised English teaching qualification in the world" and presently boasts more than 350 authorized centers in over 80 countries (Cambridge English, 2022).

Since the development of the Haycrafts' preparation course, the number of certificate programs has dramatically increased (Tanghe & Lee, 2018; Thomson, 2004) with the current spiraling popularity of online

[3] Whereas this class was taught in England, other pre-service teacher training courses in the 1950s/60s were created in Stockholm for the British Centre and elsewhere through the Centre for British Teachers (Howatt, 1984).

courses attributed to the advent of technology, projected growth of online education, and concurrent demand for teachers around the world (iResearch, 2018).

"Fast-Track" Teacher Certification

TEFL certificates have become the de facto entry-level credential for teaching English in the foreign language classroom. Surprisingly, however, little has been written on the range of certificates available nor their professional legitimacy (Ferguson & Donno, 2003; Hobbs, 2013). Research on TEFL certificates has primarily questioned whether a short-term course (typically around 120 hours) can adequately prepare teachers for the language classroom (Ferguson & Donno, 2003; Kanowski, 2004). This brevity limits what can be taught, the depth of instruction, and how much information participants can retain (Clark & Paran, 2007; Hobbs, 2013). Another voiced concern is the prioritization of classroom pedagogical practices over theoretical perspectives (Hobbs, 2013). It is believed that without a well-grounded understanding of theory, certificate-trained teachers may lack the ability to create an effective learning environment (Tanghe & Lee, 2018). This critique in ELT reflects broader research on fast-track teacher education preparation programs (e.g., Teach for America) that typically prepare predominantly middle-class, White, monolingual teachers from racially homogeneous suburban or small-town communities to teach in low-income, under-resourced urban schools (Irizarry & Donaldson, 2012). These "boot camp" style teacher models exist in a variety of forms and have grown in a market-driven economy where course providers feel increased pressure to generate revenue and compete to offer the fastest, easiest, most flexible path to teacher certification (Nygreen et al., 2015).

Short-term course proponents argue that TEFL certificates equip novice teachers with the fundamental skills needed to teach effectively (Brandt, 2006). Responding to Gibson Ferguson and Sarah Donno (2003), Sally Macpherson (2003) defended certificate courses citing that they succeed in their preparatory goal and are not intended as a terminal credential but a springboard for further professional development.

Although this reality has been questioned (see Hobbs, 2013), TEFL certificates have had a significant impact on the growth of the ELT industry, benefited teachers and their students (Kanowski, 2004; Tanghe & Lee, 2018), and created informed advocates for the educational enterprise (Labaree, 2010). TEFL certificates evidence a recognition for the preparation needed to teach English (Thomson, 2004) and dispel myths that being a "native speaker" is enough to become an effective language instructor (Phillipson, 1992).

Previous research on TEFL certificates has predominately looked at well-known in-person courses such as the Cambridge CELTA or Trinity College London CertTESOL (e.g., Mackenzie, 2019). The continued demand for teachers and acceptance of these courses as the minimum level of professional preparation has, however, motivated the growth of independent for-profit programs, resulting in wide parameters for what constitutes as a TEFL certificate (Tanghe & Lee, 2018). With no official curricular standards, accreditation process, or supervisory system, these programs have been met with apprehension (Watt & Taplin, 1997). Concerns regarding educational quality continue to persist within the current climate that has seen the growth of "budget" online programs (Tanghe & Lee, 2018). This chapter responds to the call for more research on this trend (Ferguson & Donno, 2003). In particular, it considers the emerging online TEFL certificate industry that is catering, among others, to TEFL volunteer tourists.

Becoming "TEFL Certified" Within the Context of Volunteer Tourism

The digital evolution of education and pressure to "upskill" volunteers has opened a new market to TEFL certificate providers in which profit motivated interests threaten the decommodified paradigms that lay at the ideological foundation of volunteer tourism (Mostafanezhad, 2013) and the ELT industry (Stainton, 2018b). Professionally rigorous, relevant, and effective pre-service teacher education that includes a critical and culturally sustaining pedagogical approach has the potential to support volunteers, positively impact those being served, and stand in stark contrast to a belief that the time/effort needed to qualify teachers would "place impossible constraints on volunteer recruitment" (Kenning, 2009, p. 37). At the same time, the adoption of market-based principles (Nygreen et al., 2015) and "creeping for-profit ethos" of the education sector (Cox, 2003, p. 21) risk undermining potential benefits of volunteer tourism (Wearing, 2001), reducing TEFL certificates into merely an "add on" to an already commoditized service trip.

In order to address the issues outlined above, this chapter employs TESOL International Association's Standards for Short-Term TEFL/TESL Certificate Programs to examine the management, curriculum, and potential learning outcomes of a group of TEFL certificates offered through Groupon's global e-commerce marketplace. After this evaluation, the chapter concludes with a discussion of commodification within volunteer tourism and its impact.

Methodology

This chapter draws on data gleaned from five TEFL certificates offered through www.groupon.com. The website is a popular search result for TEFL courses and mentioned *ad nauseam* on ELT Facebook pages, subreddits, blogs, and online forums such as Dave's ESL Cafe. As "one of the fastest-growing companies in history," Groupon benefits customers by offering discounted items and does not require merchants to pay any upfront advertising fees (Shontell & Goldman, 2011). The certificates on the site are used as a yardstick from which other (often more expensive) online TEFL programs differentiate themselves and are known collectively by netizens as a "Groupon TEFL."

Organizations on Groupon offering TEFL certificates are advertised under the website's language courses section. Among the eight initial TESOL/TEFL courses considered for this study, three were eliminated from the sample because their completion did not award a certificate. This resulted in five "Best of Groupon"[4] programs being chosen:

[4] A Groupon award based on course popularity/customer satisfaction.

International Open Academy (IOA), Mary Kay Jenny (MKJ), TEFL Fullcircle, Global Language Training (GLT), and Learn TEFL. These mostly US- and UK-based businesses[5] have a high probability of being recommended by volunteer tourism sponsoring organizations given their visibility online, low cost, and opportunities for affiliate marketing (e.g., International Open Academy, 2022). Certificate programs, themselves, also advertise to volunteer tourists. GLT, for instance, advertises with their "internationally accredited and accepted Global TEFL/TESOL … You will find employment in schools, collages [*sic*], universities, languages school [*sic*], business corporations, private tutoring and volunteer work around in the world" (Global Language Training, n.d., p. 15).

This chapter examines course information provided by organizations on their websites and through their affiliated Groupon pages. These data were supplemented by requests for clarification from the organizations, online student reviews, and anecdotal evidence drawn from the author's experience taking three similar courses from 2017 to 2019. A content analysis was then conducted using predefined categories drawn from TESOL International Association's[6] Standards for Short-Term TEFL/TESL Certificate Programs. This deductive approach sought to qualitatively examine Groupon TEFL courses to determine whether their content met TESOL's standards (Hsieh & Shannon, 2005). Based on preexisting knowledge and theory in the field of ELT, these criteria were drafted by TESOL's Accreditation Standards Expert Group in response to the growth of short-term TESOL certificate programs and the demand for their accountability (Staehr Fenner & Snyder, 2014). They exemplify discipline-agreed upon standards, are applicable to any short-term certificate program (i.e., onsite, hybrid, or online), and are divided into three complementary strands: Organization and Program Management Standards, Curriculum and Instruction Standards, and Candidate Standards (see Fig. 3.1).

[5] Groupon is primarily available in Europe and North American.

[6] One of the largest ELT professional associations in the world.

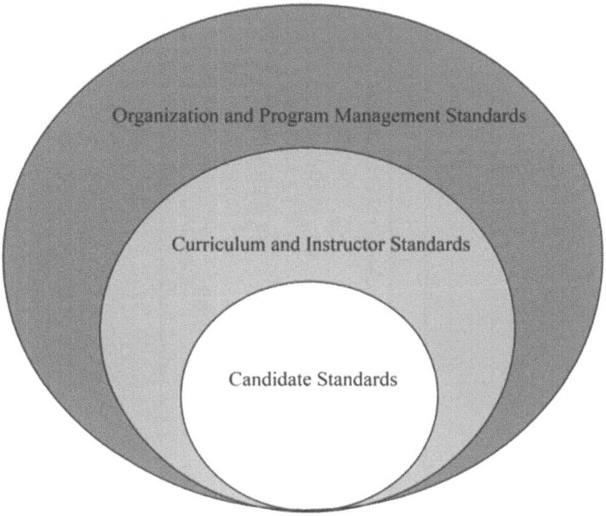

Fig. 3.1 Standards for Short-Term TEFL/TESL Certificate Programs. (Adapted from Stroupe et al., 2015)

Findings and Discussion

TESOL's Organization and Program Management Standards as well as their Curriculum and Instructor Standards provide an administrative framework to help ensure Candidate Standards are met. Each standard consists of a set of guiding domains.

Organization and Program Management Standards

TESOL's outer circle provides program management guidelines and consists of four domains: (1) mission statement, (2) program length and structure, (3) administration, and (4) candidate services (see Table 3.1).

Mission Statement

Employment is a common motivation for those pursuing a TEFL certificate, and programs markedly addressed this goal in their missions,

Table 3.1 Organization and Program Management Standards

	IOA	MKJ	TEFL Fullcircle	GLT	Learn TEFL
Mission Statement	Career focused	Career focused	TEFL career focused	TEFL career focused	TEFL career focused
Program Length and Structure	120 h 10 modules	N/A 22 modules	160 h 26 modules	120 h 6 modules	150 h 12 modules
Admin.	Open enrollment	Open enrollment	Age/language requirement	Age/language requirement	Language requirement
Candidate Services	N/A	Career advising	Career advising	Career advising	Career advising

curricula, and services. IOA's catalogue of over 100 courses included animal science, interior design, and criminology. The organization cited through them the ability to "transform your career" and become a professional in your interested industry. MKJ similarly offered an assortment of 42 courses and advertised, "Better Career Opportunities through Education." Whereas IOA and MKJ had a broad work-focused mission, GLT, TEFL Fullcircle, and Learn TEFL (as their names suggest) were narrowly focused on ELT and offered various specialist courses (e.g., teaching business English). This course diversity allows organizations to attract students with different interests and needs. It also reflects a market-infused approach to education where knowledge becomes an instrument to achieve employment (Schwartzman, 2013).

Program Length and Structure

Courses were structured from 6 (GLT) to 26 (TEFL Fullcircle) modules and subdivided into units. All programs (excluding MKJ) stated their instructional hours and met TESOL's 100–160 hour recommendation. This finding reflects a common employer requirement (Tanghe & Lee, 2018). While TESOL acknowledges online courses need to account for their hours differently than face-to-face programs (Staehr Fenner & Snyder, 2014), it was unclear how time was determined as each course was asynchronous. The number of hours required to complete each course will depend on a student's prior knowledge and aptitude. However, online reviews warrant a concern that the listed hours may not represent their true completion time. As David Watt and John Taplin (1997) caution, these pseudo-hours may instead be used to minimally meet accreditation/job requirements, compete with in-person programs, and therefore attenuate course effectiveness by providing only a brief overview of the field.

Administration

Enrollment was mostly open to anyone "who would like to try teaching" (TEFL Fullcircle). Confirming research on volunteer tourism's lack of required expertise (Bernstein & Woosnam, 2019), MKJ and IOA had no prerequisites and used this lack of selectivity to recruit applicants. Learn TEFL required fluent English speakers. GLT and TEFL Fullcircle similarly recommended a B2/C1 level[7] and had an age restriction of 18. These policies are in opposition to David Watt and John Taplin (1997), who argued for stringent admission standards that go beyond the ability to pay tuition and include a degree, relevant coursework, and an acceptable GPA. No program had an education requirement. Any mention of a degree was not related to the ability to complete the course but rather the potential for employment. Learn TEFL wrote, "Although a degree would be preferable, there are plenty of employers who will accept you without a degree." TEFL Fullcircle likewise noted, "Many countries require a degree for the work visa but there are also many countries that do not require a degree" and (ignoring employment visa requirements) incorrectly cited China as a possible employment destination.

The online marketing of "Groupon TEFL certificates" relied heavily on half-truths and embellished the degree of preparedness students receive. Programs referred to their courses as "the best available" (GLT) and boasted "the opportunity to obtain the skills, confidence and qualifications to become the best and most recognized TEFL teachers around the world" (Learn TEFL). These claims were further accentuated through various sales techniques (e.g., affiliate marketing and Groupon discounts). Other strategies such as social proofing (e.g., providing enrollment numbers and student testimonials), citing accreditation, and claiming international recognition were used to attract students and evidence program quality/credibility. GLT's "Best for Jobs" TEFL course advertised, "Your ability to speak English is a commodity in every corner of the world." TEFL Fullcircle elaborated, "In essence, the TEFL Course will give you instant access to thousands of jobs worldwide and the opportunity to live and work in another country" and made unsubstantiated claims regarding "excellent salaries" and "unrivalled travel opportunities."

Unlike employers in the Global South who want to dissuade job applicants from viewing their position as a form of leisure (Bernstein & Woosnam, 2019), programs used common tourism tropes to incentivize

---

[7] Within the Common European Framework of Reference for Languages (CEFR), B2/C1 indicate an intermediate/upper-intermediate English level.

applicants. This reflects the appeal to leisure found in volunteer tourism promotional materials and illustrates how the TEFL certificate industry is "drawing in" potential volunteer tourists whose mobilities are also linked to travel and escapism (Keese, 2011). Jobs were referenced as "adventures" (IOA), "life changing experiences" (Learn TEFL), and locations were framed as "exotic" (GLT). Learn TEFL cited their mission was "to enable others to have the same life changing experiences that we have had teaching English abroad" and moved from exaggeration to hyperbole: "Students simply sign up for a certification course, and upon completion may begin looking for jobs in far-flung locales where English teachers are highly sought, such as Asia, the Middle East, and the moon." These overstated claims mirror those reported by Ron Thomson (2004), who cautioned prospective teachers from paying for inadequate preparation when the principal ELT qualification was not a certificate but a diploma. The lack of admission requirements also reflects David Watt and John Taplin's (1997) concern for TEFL programs motivated by financial opportunism which in this study extended to untruthful and exaggerated claims.

Candidate Services

Career advising was the main service. GLT, for instance, provided "Free lifetime TEFL job placement" and considered this offering "integral" to their services. This finding is in line with Hayley Stainton (2018b), who found inexperienced TEFL tourists tended to desire the assistance and security of a placement agency. The commodified packaged TEFL product is therefore likely to appeal to young volunteer tourists who are career conscious and new to teaching.

Taking advantage of potential candidate urgency, the ability to begin classes immediately and receive instant feedback was also highlighted. This convenience was monetized. TEFL Fullcircle required applicants to wait ten business days to receive their certificate, which could be expedited for an additional fee. A certificate hard copy (a common employer requirement) necessitated an additional payment. Indicating what was truly valued (i.e., possession of a certificate), in some instances, a printed certificate was more expensive than the TEFL course itself. Other (revenue generating) services included extending program time, faster grading, and purchasing additional specialty certificates or lesson plans. Career advising therefore acted as a student incentive and an additional revenue stream for programs acting as agents. Services were intrinsically linked to financial benefits and did not necessarily improve certificate quality.

Curriculum and Instructor Standards

TESOL's Curriculum and Instructor Standards provide guidelines for curriculum development and instructor selection. They consist of two domains: (1) Curriculum, Instruction, and Assessment and (2) Instructor (see Table 3.2).

Curriculum, Instruction, and Assessment

While each certificate included similar content, some diversity existed. TEFL Fullcircle and GLT had a module on teaching English to young learners, and consistent with their career focus, IOA and Learn TEFL included a job-focused module. Instructional approaches of these lessons varied little and were primarily text-based. This lack of variety does not take into account a candidate's first language, culture, or learning style. Because volunteer tourism programs operate in numerous instructional contexts (e.g., after-school youth programs, adult education centers), it is also questionable whether coursework can anticipate the various environments in which volunteer tourists work. The common pedagogical emphasis on communicative language teaching found in certificates may additionally prove unsuitable for large classes or have varied results depending on students' sociocultural characteristics, learning goals, and/or local educational norms (Copland et al., 2016).

Multiple choice quizzes were the main form of assessment. They were used by IOA and MKJ as a form of summative assessment at the end of each module to determine a student's final grade. Learn TEFL mixed formative and summative assessment techniques including a quiz at the end of each lesson, three written assignments, and a final exam. TEFL

Table 3.2 Curriculum and Instructor Standards

	IOA	MKJ	TEFL Fullcircle	GLT	Learn TEFL
Curriculum, Instruction, and Assessment	No practicum	No practicum	No practicum	No practicum	No practicum
	Text-based	Text-based	Text-based	Text-based and videos	Text-based and videos
	Summative assessment	Summative assessment	Summative and formative assessment	Summative and formative assessment	Summative and formative assessment
Instructor	Customer support	Customer support	Customer support	Tutor support	Tutor support

Fullcircle similarly employed quizzes throughout their course followed by a summative "End of Course Assessment," which included a lesson plan, final quiz, and essay. GLT's course consisted of a variety of formative assignments (e.g., quizzes and videos) and three graded summative assignments spread throughout the course: a grammar exam, final exam, and lesson plan. Feedback on individual performance from these assessments was either automatic (i.e., computer-graded) or took between two and ten days for written assignments. No program had a practicum. This curriculum, and lack of rigorous assessment, echoes previous concerns regarding for-profit organizations and the need to maximize their returns on investment through a high volume of students and low overhead (Thomson, 2004; Watt & Taplin, 1997).

Instructor

Limited trainer information likely indicates a negligible amount of human-to-human "customer support" for most programs and, as found by Ron Thomson (2004), limited professional legitimacy. MKJ and TEFL Fullcircle stated that their learning materials were designed by industry professionals but provided no personnel information. Similarly, Learn TEFL listed personalized assistance from experienced TEFL professionals under their "key features" but lacked transparency on their trainer qualifications. IOA listed 57,201 enrolled students and "the opportunity to submit lesson plans to experienced TESOL teachers, gaining valuable feedback from experts" but only provided information on one instructor. Further, on contacting the organization it was found no lesson plans could be submitted (personal communication, May 2019). GLT was the notable difference, providing details on 11 of their experienced tutors. However, with more than 60,000 program graduates, it is questionable whether individual attention and personalized input is possible. This absence of support and no supervised practicum likely translates to inadequately prepared certificate graduates without the autonomy required to generate their own pedagogical knowledge (Kumaravadivelu, 2003) or ability to distance themselves from fixed techniques associated with Westernized approaches to language instruction (Thomson, 2004).

Self-paced and lacking real-time human interaction, certificates were packaged products available on an online market divorced from their unavowed creator (Chau, 2010). Curriculum and instruction resembled Massive Open Online Courses (MOOCs), a fashionable model of education delivery, which have been criticized as putting students (i.e., customers) to work by substituting human-to-human interactive elements such as direct contact between teachers and students with more efficient/predictable digital content (e.g., prepackaged/uniform modules) to maximize returns (Ritzer, 1993).

Candidate Standards

Candidate Standards are composed of the foundational knowledge and skills that should be possessed on certificate completion. They represent the core goal of the program in that they support candidates in their teaching role and consist of the following: (1) language, (2) culture, (3) instruction and assessment, and (4) professionalism (see Table 3.3).

Language

Only IOA provided a dedicated module on linguistics. Given the language requirement for some programs and their primarily Western clientele, it was likely felt enrolled candidates possessed a sufficient level of language awareness. TEFL Fullcircle advertised "translate your fluency in English into effective classroom instruction" and did not have a linguistics module but instead instructed candidates on how to teach phonology. Similarly, whereas GLT and TEFL Fullcircle reviewed grammar, MKJ and Learn TEFL had a module on teaching grammar.

Assuming TEFL certificate students possess intuitive language fluency and not addressing explicit language awareness may qualify individuals to teach English that do not have a full grasp of the language (Thomson, 2004). During data collection this concern was validated on various ELT message boards where individuals offered to take online TEFL courses for those who could not pass, and "TEFL qualified" teachers who did not appear to have sufficient language ability/awareness inquired about jobs.

Table 3.3 Candidate Standards

	IOA	MKJ	TEFL Fullcircle	GLT	Learn TEFL
Language	Linguistics module	Teaching grammar	Teaching phonology and grammar module	Grammar unit	Teaching grammar
Culture	2 culture units	2 culture modules	N/A	N/A	N/A
Instruction and Assessment	Practical elements	Practical elements	Practical elements	Practical elements	Practical elements
Professionalism	Price: $5 Accredited (ICOES)	Price: $29 Accredited (OTEA)	Price: $39 Accredited (ACCREDITAT)	Price: $29 Accredited (ACTDEC)	Price: $39 Accredited (IARC)

Certificates have also been critiqued as privileging "native English speakers" over more qualified "non-native English speakers" by implicitly equating language usage competency with language teaching competency and playing an unhelpful role in constructing the reputation of the profession as easy to pursue, provided one is a native English speaker (Ferguson & Donno, 2003). The assumption that students from the Global North possess the necessary descriptive knowledge of English reflects the fallacious notion that the ideal teacher of English is a native speaker (Phillipson, 1992) and has contributed to unjust racial hierarchies in ELT that are normalized and reproduced through teacher education (Anya, 2018). This lack of explicit instruction is especially concerning when volunteer tourism sponsors suggest volunteers are needed for their language knowledge, innovative pedagogies, and instructional effectiveness (see Jakubiak, this volume).

Culture

Two programs loosely addressed the connections among language, culture, and learning. MKJ had a module on culture in ELT and one on managing a culturally diverse classroom. IOA included a unit on joining a different culture and one entitled "Understanding Cultural Backgrounds." Preparing students to contextualize their teaching through culturally relevant or asset-based pedagogies is important within the context of English language volunteer tourism because certificate graduates, often White, young adults from homogeneous communities in the Global North, typically disperse to teach members of racially/ethnically marginalized communities in the Global South (Tanghe & Lee, 2018). It is this sociocultural and economic difference that forms part of volunteer tourism's allure (Keese, 2011) and the packaged experience of traveling abroad to a foreign location (Tomazos, 2021). In the surveyed programs, however, the importance of context was mostly overlooked and instead replaced by unsupported claims such as "once you complete this course you will have all of the skills and knowledge to teach English anywhere around the world" (GLT). This declaration reflects a technocratic view of education which "frames teaching as a set of discrete skills or 'best practices' that one can learn, and then apply in any context" (Nygreen et al., 2015, p. 118), does not recognize the need for teacher education to equip learners with intercultural skills so that they may interact effectively and appropriately in multicultural settings (Anya, 2018; Stanley & Murray, 2013), and ignores recognition among host organizations that volunteer success is not solely based on their technical skill level but requires intercultural awareness and understanding (Lough et al., 2018).

Instruction and Assessment

All courses covered classroom management and lesson planning with varying focus on specialized knowledge such as using drama in the classroom (GLT) and technology (TEFL Fullcircle). Despite the popularity of Groupon TEFL certificates for online teaching, surprisingly no course had this focus.[8] Assessment was explicitly covered by all programs except GLT, which instead included units on error correction. These findings reflect the privileging of pedagogical skills immediately applicable to the classroom over theoretical perspectives (Hobbs, 2013; Tanghe & Lee, 2018). This gap in theoretical grounding is exacerbated by no supervised teaching or classroom observation, and a potentially insufficient linguistic/pedagogical foundation making transitioning into teaching day-to-day lessons challenging. Further, policies and procedures did not ensure accepted candidates had the appropriate knowledge, experience, or skills to be successful in the program or the ability to meet their goals upon course completion.

Professionalism

No program explicitly covered professionalism. TEFL Fullcircle had a module on reflection and a unit broadly covering classroom behavior. MKJ had a module on essential teacher qualities. Likely symptomatic of the overarching work-focused mission and time constraints, focus centered on how to be a professional rather than behaving professionally.

[8] A cursory glance of the TEFL certificates offered on Groupon (post-data collection) suggests some have expanded their library to include an online teaching focus with a select few catering specifically to this market segment.

Program professionalism (i.e., legitimacy) was also emphasized. TEFL Fullcircle, for example, included "professional" in the name of each of their core certificates (e.g., Professional 120-hour Online TEFL) and advertised that on course completion students would become "professionally qualified." This is in lieu of any standard qualifying measure in ELT (Stanley & Murray, 2013). Programs also flippantly stated they were accredited, although each had a different accreditor. This lack of uniformity, and outside inspection from questionable non-governmental organizations who set their own membership rules and standards, brings into question course legitimacy.

Rather than addressing the aspects of professionalism recommended by TESOL's Short-Term Certificate Program Standards (i.e., professional learning and growth, professional ethics and behavior), the private programs surveyed in this study took a free-market approach prioritizing profit, efficiency, and customer satisfaction while treating education itself as a commercial transaction (Schwartzman, 2013). Subsequently, within the so-called unattainable triangle,[9] speed and price took precedence. Targeting good-intentioned volunteer tourists who are often young adults (Wearing, 2001) and may not have the capital to purchase more expensive training, courses ranged between \$5 and \$45. Through Groupon, for example, IOA's \$499 certificate was "99% off" and cost \$5. This is significantly cheaper than in-person courses like the CELTA which can cost over \$2,000. While price does not necessarily equate to value, an education decision based on the cheapest option and/or the need for a check mark to bypass legal restrictions is not indicative of a commitment to ELT or the pursuit of future professional development.

Conclusion

This chapter highlights a group of teaching certificates available "on the market" and several pandemic issues linked to the increasingly commoditized volunteer tourism and language learning/teaching industry. While pre-service training may not be feasible nor necessary for volunteers in all

[9] Also referred to as the "iron triangle," this maxim posits that only two can prevail among the triad of cost, time, and quality.

circumstances (see Carpenter, 2015), becoming TEFL certified has currency. Private companies have exploited this cachet, exacerbating problems linked to volunteer preparation and the ELT industry more broadly.

TEFL certificates are designed to provide prospective teachers with the fundamentals needed to be successful. Responding to an ambiguous professional milieu[10] (Bernstein, 2021; Tanner, 2003), individuals choose Groupon courses because they are inexpensive, convenient, and lead to a "recognized" qualification. However, this chapter questions the true benefit of mass marketed online TEFL certificates for those being taught (i.e., the volunteer and "voluntoured"). The programs surveyed in this study failed to sufficiently meet TESOL's standards and subsequently evidenced the continued commoditization of language education within the context of volunteer tourism. They positioned qualifying for a teaching position as cheap and easy: anyone can do it from anywhere in no time at all. This retrograde step toward ELT credibility competes with more comprehensive programs, adds additional obstacles for uninformed applicants, and further entrenches a negative professional perception of the vocation. That volunteers can now "purchase" TEFL qualifications further entrenches volunteer tourism within the commodified paradigms found in mass tourism (Wearing, 2001) and undermines the industries' "carefully cultivated self-image of decommodification" (Coren & Gray, 2012, p. 233). Consequently, students enroll in a business model where attracting a large quantity of students takes precedence over educational quality. This produces questionably prepared teachers and offers a form of veiled legitimacy for companies to market their tutors as "TEFL qualified."

While this study focused on the content and structural constraints of online TEFL certificates, their implicit message suggests that learning to teach those from the Global South is something volunteers can do with ease and efficiency. This sentiment undermines attempts to professionalize ELT, devalues local instructors, and positions the recipients of volunteer tourism as helpless and in need of foreign aid (Jenks & Lee, 2020;

[10] The imprecision found in the ELT industry extends to the deceptive marketing of TEFL certificates, which (as evidenced in this study), include overstated instructional hours, claims of "international" recognition and accreditation, as well as the promise that teachers will be equipped with the skills to find comfortable work anywhere in the world.

Labaree, 2010; Nygreen et al., 2015; Thomson, 2004). The packaging of TEFL certificates with volunteer tourism offers members of the Global North new "opportunities" to perform generosity and "save" people in disadvantaged communities by teaching them English. Volunteers can feel good without seriously considering issues of power, difference, or structural inequality (Nygreen et al., 2015) while enhancing their skills and career prospects. Despite ELT scholars advocating critical and reflective teacher approaches and new pressures/mandates calling for volunteers to be better prepared, the certificates in this study did not move students beyond "native speaker" paradigms nor provide an adequate pedagogical foundation. TEFL volunteer tourists can "do well" without personal sacrifice, risk of putting their professional reputation or self-esteem in jeopardy, and move on to their "real life" maintaining the satisfaction and prestige linked to helping disadvantaged people regardless of the effectiveness of their teaching (Labaree, 2010).

TESOL's development of a set of standards for short-term TEFL certificates is laudable in an environment with no international standardization body. Because TESOL's standards were designed as a framework for program self-assessment, some domains were beyond the scope of this study. Others were not applicable for online courses such as facility adequacy. Although TESOL recognizes that satisfying all standards may be a challenge for some programs, further accommodation for online courses is needed. Future research would benefit from using TESOL's standards to analyze other online courses as well as established courses such as the CELTA. Inquiry should also extend to accreditation organizations. While quality assurance reviews by external auditing organizations can be used to improve the effectiveness of instruction and are potentially a positive step forward (Watt & Taplin, 1997), accreditation agencies bestow a veneer of legitimacy to a training that, ultimately, may leave candidates unprepared. Further research on the demarcation between volunteerism, tourism, and work is also needed in relation to ELT. The popularity of online learning and remote volunteering (with a heightened focus on skills development) is predicated to escalate in the post COVID-19 era (Perold et al., 2021). As online education providers become undistinguishable through deceptive marketing practices, price becomes the main differentiator. The lack of clarity within ELT (Bernstein, 2021), increased automation of curriculum and instruction (Noble, 1998), and relative ease of creating a professional looking website contribute to "a race to the bottom" (Cox, 2003, p. 27) where inadequate preparation does a disservice to teachers and the communities in which they ostensibly "help" (Nygreen et al., 2015). Certificate courses must therefore be reassessed on an ongoing basis as they continue to be used as a form of quality control and global learner/teacher needs evolve.

Through advancing understanding of the present ELT landscape, this study provides input to gatekeepers, informs aspiring certificate candidates, and advocates for those ultimately impacted by volunteer tourism (i.e., the voluntoured). It documents how unequal power relations between volunteers and the voluntoured are codified through commercial TEFL certification programs marketed to volunteer tourists and a gap between what online TEFL certificates purport to do and what they actually do. As short-term certificate courses become ubiquitous and the quantitative need for teachers continues to outpace the number of qualified individuals (Thomson, 2004), prudent procedures spearheaded by individual stakeholders, professional associations, and administrative bodies are needed to combat market-influenced financial interests and advocate toward increased industry regulation/standardization both within the volunteer tourism industry and at the local, state, and federal levels.

References

Anya, U. (2018). Race and ethnicity in teacher education. In J. I. Liontas & M. DelliCarpini (Eds.), *The TESOL encyclopedia of English language teaching*. Wiley. https://doi.org/10.1002/9781118784235.eelt0673

Bernstein, J. D. (2021). Gobbledygook: The proliferation of jargon in English language teaching. *TESOL Journal, 12*, e541. https://doi.org/10.1002/tesj.541

Bernstein, J. D., & Woosnam, K. L. (2019). Same same but different: Distinguishing what it means to teach English as a foreign language within the context of volunteer tourism. *Tourism Management, 72*, 427–436. https://doi.org/10.1016/j.tourman.2019.01.010

Brandt, C. (2006). Allowing for practice: A critical issue in TESOL teacher preparation. *ELT Journal, 60*(4), 355–364. https://doi.org/10.1093/elt/ccl026

Brown, F., & Hall, D. (2008). Tourism and development in the global south: The issues. *Third World Quarterly, 29*(5), 839–849. https://doi.org/10.1080/01436590802105967

Brown, P., Hesketh, A., & Williams, S. (2003). Employability in a knowledge-driven economy. *Journal of Education and Work, 16*(2), 107–126. https://doi.org/10.1080/1363908032000070648

Brown, S., & Morrison, A. M. (2003). Expanding volunteer vacation participation: An exploratory study on the mini-mission concept. *Tourism Recreation Research, 28*(3), 73–82. https://doi.org/10.1080/02508281.2003.11081419

Cambridge English. (2022). *CELTA: Certificate in teaching English to speakers of other languages*. Retrieved December 10, 2022, from https://www.cambridgeenglish.org/teaching-english/teaching-qualifications/celta/

Carpenter, K. (2015). Childhood studies and orphanage tourism in Cambodia. *Annals of Tourism Research, 55*, 15–27. https://doi.org/10.1016/j.annals.2015.08.010

Chau, P. (2010). Online higher education commodity. *Journal of Computing in Higher Education, 22*, 177–191. https://doi.org/10.1007/s12528-010-9039-y

Clark, E., & Paran, A. (2007). The employability of non-native-speaker teachers of EFL: A UK survey. *System, 35*, 407–430. https://doi.org/10.1016/j.system.2007.05.002

Copland, F., Mann, S., & Garton, S. (2016). Introduction: Positions, experiences and reflections on the native speaker issue. In F. Copland, S. Garton, & S. Mann (Eds.), *LETs and NESTs: Voices, views and vignettes* (pp. 5–19). British Council.

Coren, N., & Gray, T. (2012). Commodification of volunteer tourism: A comparative study of volunteer tourists in Vietnam and in Thailand. *International Journal of Tourism Research, 14*, 222–234. https://doi.org/10.1002/jtr.848

Cox, A. M. (2003). None of your business: The rise of the University of Phoenix and for-profit education—And why it will fail us all. In B. Johnson, P. Kavanagh, & K. Mattson (Eds.), *Steal this university: The rise of the corporate university and the academic labor movement* (pp. 15–32). Routledge.

De Costa, P. I., Park, J. S.-Y., & Wee, L. (2016). Language learning as linguistic entrepreneurship: Implications for language education. *The Asia-Pacific Education Researcher, 25*(5–6), 695–702. https://doi.org/10.1007/s40299-016-0302-5

Ferguson, G., & Donno, S. (2003). One-month teacher training courses: Time for a change? *ELT Journal, 57*(1), 26–33. https://doi.org/10.1093/elt/57.1.26

Frilund, R. (2018). Teasing the boundaries of 'volunteer tourism': Local NGOs looking for global workforce. *Current Issues in Tourism, 21*(4), 355–368. https://doi.org/10.1080/13683500.2015.1080668

Global Language Training. (n.d.). *Brochure*. Retrieved December 10, 2022, from https://www.globaltefl.uk.com/brochure.pdf

Go Overseas. (2020). Volunteer teaching abroad programs. Retrieved May 10, 2023, from https://www.gooverseas.com/volunteer-abroad/teaching

Gray, N. J., & Campbell, L. M. (2007). A decommodified experience? Exploring aesthetic, economic and ethical values for volunteer ecotourism in Costa Rica. *Journal of Sustainable Tourism, 15*(5), 463–482. https://doi.org/10.2167/jost725.0

Guttentag, D. A. (2009). The possible negative impacts of volunteer tourism. *International Journal of Tourism Research, 11*(6), 537–551. https://doi.org/10.1002/jtr.727

Haycraft, J. (1988). The first international house preparatory course: A historical overview. In T. Duff (Ed.), *Explorations in teacher training: Problems and issues* (pp. 1–9). Longman.

Heath, S. (2007). Widening the gap: Pre-university gap years and the 'economy of experience.' *British Journal of Sociology of Education, 28*(1), 89–103. https://doi.org/10.1080/01425690600996717

Heller, M. (2010). The commodification of language. *Annual Review of Anthropology, 39*(1), 101–114. https://doi.org/10.1146/annurev.anthro.012809.104951

Hobbs, V. (2013). 'A basic starter pack': The TESOL certificate as a course in survival. *ELT Journal, 67*(2), 163–174. https://doi.org/10.1093/elt/ccs078

Howatt, A. P. R. (1984). *A history of English language teaching*. Oxford University Press.

Hsieh, H.-F., & Shannon, S. E. (2005). Three approaches to qualitative content analysis. *Qualitative Health Research, 15*(9), 1277–1288. https://doi.org/10.1177/10497323052766

International House London. (n.d.). *About IH London*. Retrieved May 10, 2023, from https://www.ihlondon.com/why-us/about-ih-london/

International Open Academy. (2022). *Join the affiliate program*. Retrieved May 10, 2023, from https://internationalopenacademy.com/pages/affiliates

International Volunteer HQ. (n.d.). *Teaching Volunteer Abroad Programs*. Retrieved May 10, 2023, from https://www.volunteerhq.org/volunteer-abroad-projects/teaching/

iResearch. (2018). *China's Online Education Sector Data in 2017*. www.iresearchchina.com/content/details7_40642.html

Irizarry, J., & Donaldson, M. L. (2012). Teach for América: The Latinization of U.S. schools and the critical shortage of Latina/o teachers. *American Educational Research Journal, 49*(1), 155–194. https://doi.org/10.3102/0002831211434764

Jakubiak, C. (2016). Ambiguous aims: English-language voluntourism as development. *Journal of Language, Identity & Education, 15*(4), 245–258. https://doi.org/10.1080/15348458.2016.1195270

Jakubiak, C. (2020). "English is out there—You have to get with the program": Linguistic instrumentalism, global citizenship education, and English-language voluntourism. *Anthropology & Education Quarterly, 51*(2), 212–232. https://doi.org/10.1111/aeq.12332

Jenks, C. J., & Lee, J. W. (2020). Native speaker saviorism: A racialized teaching ideology. *Critical Inquiry in Language Studies, 17*(3), 186–205. https://doi.org/10.1080/15427587.2019.1664904

Kanowski, S. (2004). Helter CELTA: Do short courses equal 'best practice' in teacher training? *TESOL in Context, 13*(2), 21–27.

Keese, J. R. (2011). The geography of volunteer tourism: Place matters. *Tourism Geographies, 13*(2), 257–279. https://doi.org/10.1080/14616688.2011.567293

Kenning, M.-M. (2009). Globalization and the use of volunteers in ELT: Enhancing volunteer impact. *CamTESOL Conference on English Language Teaching: Selected Papers, 5*, 36–44. https://camtesol.org/Download/Earlier_Publications/Selected_Papers_Vol.5_2009.pdf

Kumaravadivelu, B. (2003). *Beyond methods: Macrostrategies for language teaching*. Yale University Press.

Labaree, D. (2010). Teach for America and teacher ed: Heads they win, tails we lose. *Journal of Teacher Education, 61*(1–2), 48–55. https://doi.org/10.1177/0022487109347317

Lough, B. J., Tiessen, R., & Lasker, J. N. (2018). Effective practices of international volunteering for health: Perspectives from partner organizations. *Globalization and Health, 14*(1), 1–11. https://doi.org/10.1186/s12992-018-0329-x

Lyons, K., Hanley, J., Wearing, S., & Neil, J. (2012). Gap year volunteer tourism: Myths of global citizenship? *Annals of Tourism Research, 39*(1), 361–378. https://doi.org/10.1016/j.annals.2011.04.016

Mackenzie, L. (2019). Investigating reflection in written assignments on CELTA courses. *ELT Journal, 73*(1), 11–20. https://doi.org/10.1093/elt/ccy037

Macpherson, S. (2003). The short intensive teacher-training course. *ELT Journal, 57*(3), 297–300. https://doi.org/10.1093/elt/57.3.297

McMillon, B., Cutchins, D., & Geissinger, A. (2012). *Volunteer vacations: Short-term adventures that will benefit you and others* (11th ed.). Chicago Review Press.
Mostafanezhad, M. (2013). The politics of aesthetics in volunteer tourism. *Annals of Tourism Research, 43*, 150–169. https://doi.org/10.1016/j.annals.2013.05.002
Noble, D. F. (1998). Digital diploma mills: The automation of higher education. *Science as Culture, 7*(3), 355–368. https://doi.org/10.1080/09505439809526510
Nygreen, K., Madeloni, B., & Cannon, J. (2015). "Boot camp" teacher certification and neoliberal education reform. In K. M. Sturges (Ed.), *Neoliberalizing educational reform: America's quest for profitable market-colonies and the undoing of public good* (pp. 101–121). Sense Publishers. https://doi.org/10.1007/978-94-6209-977-7_5
Palacios, C. M. (2010). Volunteer tourism, development and education in a postcolonial world: Conceiving global connections beyond aid. *Journal of Sustainable Tourism, 18*(7), 861–878. https://doi.org/10.1080/09669581003782739
Park, J. S.-Y. (2021). *In pursuit of English: Language and subjectivity in neoliberal South Korea*. Oxford University Press.
Perold, H., Allum, C., Lough, B. J., & Mati, J. M. (2021). *COVID-19 and the future of volunteering for development*. International forum for volunteering in development (forum).. https://forum-ids.org/wp-content/uploads/2016/03/Forum-Survey-Report-COVID-19-and-the-Future-of-Volunteering-for-Development-1.pdf
Phillipson, R. (1992). *Linguistic imperialism*. Oxford University Press.
Raymond, E. (2008). "Make a difference!": The role of sending organizations in volunteer tourism. In K. D. Lyons & S. Wearing (Eds.), *Journeys of discovery in volunteer tourism: International case study perspectives* (pp. 48–60). CABI.
Raymond, E. M., & Hall, C. M. (2008). The development of cross-cultural (mis)understanding through volunteer tourism. *Journal of Sustainable Tourism, 16*(5), 530–543. https://doi.org/10.1080/09669580802159610
Ritzer, G. (1993). *The McDonaldization of society*. Pine Forge Press.
Schwartzman, R. (2013). Consequences of commodifying education. *Academic Exchange Quarterly, 17*(3), 41–46.
Shontell, A., & Goldman. L. (2011, June 4). *GROUPON: From the ashes of a dead startup to a billion-dollar company in 2 years*. https://www.businessinsider.com/groupon-from-the-ashes-of-a-dead-startup-to-a-billion-dollar-company-in-2-years-2011-6

Simpson, K. (2004). 'Doing development': The gap year, volunteer-tourists and a popular practice of development. *Journal of International Development, 16*, 681–692. https://doi.org/10.1002/jid.1120

Sin, H. L. (2010). Who are we responsible to? *Locals' tales of volunteer tourism. Geoforum, 41*(6), 983–992. https://doi.org/10.1016/j.geoforum.2010.08.007

Sinervo, A. (2017). Interdependent industries and ethical dilemmas NGOs and volunteer tourism in Cusco, Peru. In A. Lashaw, C. Vannier, & S. Sampson (Eds.), *Cultures of doing good: Anthropologists and NGOs* (pp. 142–162). University of Alabama Press.

Staehr Fenner, D., & Snyder, S. (2014). TESOL International Association ASEG Draft TESOL Short-Term Certificate Program Standards.. https://www.tesol.org/docs/default-source/press-releases/tesol-short-term-certificate-standards-public-review.pdf?sfvrsn=4

Stainton, H. (2018a). TEFL tourism: The tourist who teaches. *Tourism Geographies, 20*(1), 127–143. https://doi.org/10.1080/14616688.2017.1298151

Stainton, H. (2018b). The commodification of English language teaching in tourism: A sustainable solution? *Tourism Management Perspectives, 25*, 123–130. https://doi.org/10.1016/j.tmp.2017.11.021

Stanley, P., & Murray, N. (2013). 'Qualified'? A framework for comparing ELT teacher preparation courses. *Australian Review of Applied Linguistics, 36*(1), 102–115. https://doi.org/10.1075/aral.36.1.05sta

Stroupe, R., Dormer, J., Giraldez, R., Kuhlman, N., & Mazloom-Farzaghy, R. (2015). *TESOL introduces standards for short-term TEFL/TESL certificate programs [Webinar]*. TESOL International Association. http://eventcenter.commpartners.com/se/Meetings/Playback.aspx?meeting.id=262659

Tanghe, S., & Lee, A. S.-J. (2018). TEFL (teaching English as a foreign language) certificates. In J. I. Liontas & M. DelliCarpini (Eds.), *The TESOL encyclopedia of English language teaching*. Wiley. https://doi.org/10.1002/9781118784235.eelt0271

Tanner, M. W. (2003). Tips for facilitating full-time employment in TESOL. *TESOL Journal, 12*(2), 40–42. https://doi.org/10.1002/j.1949-3533.2003.tb00131.x

TEFL Institute of Ireland. (2022). Our top 3 courses. Retrieved May 10, 2023, from www.tefl.ie

Thomson, R. I. (2004). Buyer beware: Professional preparation and TESL certificate programs in Canada [special issue]. *TESL Canada Journal, 4*, 40–57. https://doi.org/10.18806/tesl.v0i0.1039

Thornbury, S. (2001). The unbearable lightness of EFL. *ELT Journal, 55*(4), 391–396.

Tomazos, K. (2021). Ethics of volunteering in tourism: Ethics of the heart. In K. Holmes, L. Lockstone-Binney, K. A. Smith, & R. Shipway (Eds.), *The Routledge handbook of volunteering in events, sport and tourism* (pp. 375–388). Routledge.

Watt, D., & Taplin, J. (1997). The least one should expect of TESL/TEFL programs. *TESL Canada Journal, 14*(2), 72–74. https://doi.org/10.18806/tesl.v14i2.687

Wearing, S. (2001). *Volunteer tourism: Experiences that make a difference*. CABI.

Wearing, S., & McGehee, N. G. (2013). Volunteer tourism: A review. *Tourism Management, 38*, 120–130. https://doi.org/10.1016/j.tourman.2013.03.002

Wearing, S., Young, T., & Everingham, P. (2017). Evaluating volunteer tourism: Has it made a difference? *Tourism Recreation Research, 42*(4), 512–521. https://doi.org/10.1080/02508281.2017.1345470

4

Translating the Value of Global Languages: Learning/Teaching Spanish/English Within Volunteer Tourism in Cusco, Peru

Aviva Sinervo

Negotiating Linguistic Practice

Sitting alone on a bench in the main plaza of touristic Cusco, I would inevitably be approached by children vending a variety of souvenirs and handicrafts. I had my choice of alpaca sweaters, yarn-knitted finger puppets, oil paintings, and postcards depicting landscapes and peoples of the Andean Sacred Valley region—and I also seemingly had my choice of the language we used to converse. I was predominantly approached in English: "Hello Miss, how are you doing? Would you like to look? Would you like to buy?" I often responded in Spanish to show my interest in the product and in the vendor, since I was researching children's selling strategies and participation in household economies. Most of the children

A. Sinervo (✉)
San Francisco State University, San Francisco, CA, USA
e-mail: asinervo@sfsu.edu

L. S. Schedel, C. Jakubiak (eds.), *Voluntourism and Language Learning/Teaching*, Palgrave Advances in Language and Linguistics,
https://doi.org/10.1007/978-3-031-40813-7_4

continued their part of the dialogue in English, using information they gathered about me (Where are you from? How long are you here?) to extend the interaction, to share facts they knew about my country (Oh, the U.S.! Your capital is Washington and your president is …), and to grow their own knowledge. I did not have my choice of the language of conversation after all, as we parried back and forth between English and Spanish, deepening our mutual understanding and acquaintance through our negotiated forms of linguistic practice.

I am not unique as a foreigner in using conversations with child vendors to practice my Spanish-speaking skills. Language teachers, who work at the dozen schools catering to short- and long-term tourists, explain that Peru is well-known for its "quality Spanish": pronunciation that is clear, conjugations that are less complicated, speaking-speeds that are leisurely, and vocabulary with fewer slangs and colloquialisms. As one of the well-known "Spanishtowns" on the "gringo trail" (Stanley, 2016), tourists spend time in Cusco in order to take language classes, and they do their oral discourse homework in the streets and NGOs (non-governmental organization) where they also volunteer.

Yet local child vendors, many of whom attend NGO afterschool projects staffed by foreign tourists, are interested in practicing their own English. In a city like Cusco, where the majority of current and future livelihood paths lead through the tourism industry, it is imperative to speak this global language comfortably. Many young vendors claim that they are motivated to sell for English dialogue opportunities as much as for a wage: a successful encounter could be measured not just by the sale but also by the quality of the interaction. English, after all, is a hegemonic language that forcefully crosses borders through cosmopolitan desires and within political economies; its value is undisputed in realms of international travel and business (Crystal, 2012).

Even as many tourists agree with the value of English and view it as a resource they have to share when teaching children in afterschool programs, tourists also premise the goals for their trip on the growing global value of Spanish. Mobile, cosmopolitan, and middle-/upper-class native English speakers highly value multilingualism, which is itself a cultural commodity (Heller, 2010; O'Rourke & DePalma, 2017). Such "linguistic entrepreneurship"—focused on the importance of using "language-related resources for enhancing one's worth in the world"—considers the affective, moral, and economic dimensions of self-development (De Costa et al., 2021, p. 139), in sum positioning language learning as a logical endeavor to consolidate into leisure experiences such as travel. Child vendors likewise perceive the benefits to their communities when tourists strengthen Spanish fluency. Tourists who are Spanish speakers are more interesting to talk with, more invested in the places and cultures they visit, and more likely to return to Cusco and recommend it to their networks. Cusco's reputation is enhanced as not only an interesting cultural place but also a good destination to practice language, make a difference, and possibly make new friends across boundaries of age and culture.

This chapter uses interactions between local children and volunteer tourists to consider the negotiated value of the global languages of Spanish and English. Bidirectional language exchange facilitates intercultural learning (Stanley, 2020) as tourists from Europe, North America, and Australia enroll in Spanish language schools that provide volunteer opportunities in afterschool programs catering to *ambulante* (mobile) child vendors (Sinervo, 2011b, 2015, 2017). As such immersion travel experiences have become an increasingly popular way to buttress a resume or fill a gap year, the promise of multilingualism (Schedel, 2023) has been sweetened by coupling formal language classes with volunteer tourism placements. Tourists leverage engagement with Spanish-speaking children and aid workers to facilitate their own deeper linguistic immersion as well as meaningful connections with locals, a desired byproduct of an authentic tourist experience (Huberman, 2012). Meanwhile, child vendor program attendees appreciate conversing with tourists to practice English, a necessary skill to facilitate potential future entry into more formal sectors of the tourism industry such as restaurants, hotels, and travel agencies. Tourists also seek to teach children English as a way to do meaningful volunteering work that makes a difference to local communities.[1]

Nevertheless, language exchange is fraught with tensions and power struggles, as children and tourists differentially value English versus Spanish, and vacillate between being givers and receivers—teachers and

[1] On "making a difference" through volunteer tourism, among many others, see Butcher and Smith (2015); Mostafanezhad (2014); and Wearing (2001).

learners—of language instruction and/or practice opportunities. Instruction through a class and practice through a conversation are contested in their weighted utility, especially when one party wants to formally teach and the other wants to build relationality through discussion (see Doerr, this volume). In the entanglement of volunteer and language tourisms, we see the importance of both languages as highly relevant locally and globally, and we can understand the participation in volunteer tourism (as either tourist volunteer or child aid recipient) as a vehicle for the languages' mutual commodification.

Below I share background on Cusco as a touristic space, unpacking changes in the tourism industry and charting the co-emergence of language and volunteer tourisms alongside neoliberal shifts in resource provision though NGOs. I position children as important actors influencing this climate of encounter across differences in nationality, class, age, language, and forms of mobility. This section concludes with an overview of the scope of my fieldwork methodologies.

In the main body of the chapter, I analyze three ethnographic portraits that highlight tourist volunteer and child worker negotiations over language teaching and learning. In the first, I problematize how fleeting and unstable curricular goals and methods emerge from the logistical and financial dimensions of short-term volunteerism. Here we see that the aspired generative value of merged Spanish language learning and volunteering is limited by the itinerant temporal structures of tourism. I next consider how English is used among children to enable what I call "linguistic dreaming." English is the language of mobility—both physical and psychological, socioeconomic and symbolic—and children strategically repurpose English-learning sessions to express their desires. I end this section by proposing that the value of volunteer tourist-facilitated language learning is diminished when positioned through school-like "enrichment" rather than as foundational and relational conversation: a view that is especially salient when considering compulsory public school English education that takes place in Peru. This prefaces my later analysis on how formal language training can actually marginalize and discount the value of colloquial engagement, delegitimizing children as agentive learners.

I use the second case study to build on the contradictions that emerged earlier in the inherent "meaningfulness" of volunteers teaching English without training or support, and often without ready cooperation and desire from the children who are supposed to be learning. I deepen my analysis of why children might both appeal for language training and resist it, and why tourists might seek to be English language teachers (see Stainton, 2018) and feel most comfortable structuring their lessons around grammar instruction. Here we consider language teaching as a form of "hard work" that can also be "rewarding" because it is viewed as creating value. Nevertheless, what does it mean for these teaching and learning contexts that children often view English formal instruction as a waste of time? If children would rather engage in conversational practice, what does that tell us about how English is best learned and who gets to control that learning? When we add that tourists also interact with children to facilitate tourists' own language practice in Spanish, we can see the possibilities inherent, but perhaps unharnessed, in existing exchanges.

Building on the emotional and imaginative qualities of language, the third and final vignette demonstrates how even basic Spanish proficiency is used to enable affective connections between tourists and local children. English is the language of possibility, yet Spanish is the language of trust and friendship. Moreover, children are not just learners of English but also teachers of Spanish—a point made earlier. This final section draws our attention to how both children and volunteers already possess multilingual resources that can be used for meaningful work and learning, *and* profound connections across differences. I emphasize that linguistic cosmopolitanisms exist in the spaces between instruction and implementation: perhaps the spaces where friendship trumps the inherent power inequalities between teacher and learner. As I conclude by returning to my opening encounters with street vendors, I note that both groups in this tripartite volunteer-language-vending tourisms nexus want to be both teacher and learner.

Throughout I shift back and forth between local children's and international tourists' perspectives, while tracking a few connective conclusions underpinning all the data. The value of teaching and learning Spanish and English in these exchanges occupies an uneasy space between professionalism and informality. While vendors claim that they want higher-quality English developmental support, they also reject the classroom as the place to grow their language skills. In the end, I puzzle through the value of different models for learning how to use language when its intention is for street encounters and friendly dialogues.

Where Locals, Tourists, and Anthropologists Meet

Evolving Tourism Industries: Cusco from 1980s to 2010s

Cusco, Peru, as the gateway city to Machu Picchu and the Sacred Valley of the Incas, has experienced tourist development at a rapid pace throughout the last three decades.[2] With UNESCO World Heritage site designation in the early 1980s (Silverman, 2002; UNESCO, 2023) and the end of guerilla and government hostilities in the mid-1990s (Stern, 1998), foreigners began slowly returning to the city. Tourists of widely varying ages, intentions, and national and socioeconomic backgrounds come to this ancient yet increasingly cosmopolitan mecca to tour archaeological sites, participate in spiritual and adventure tourism (Hill, 2007), and experience Peruvian cuisine (Hall, 2019).

The late 1990s and especially into the early 2000s saw two new forms of tourism take hold: language tourism and volunteer tourism. The rise of these forms of tourism globally have been well-documented,[3] and Peru's reputation for "clean" and easy-to-understand Spanish made it a desirable location to learn through the combined approach of formal classes and host-family immersions. The pitch for language tourism locally was established by business savvy entrepreneurs, who leveraged international networks to form study abroad opportunities advertised

---

[2] Tragically, the COVID-19 pandemic of 2020 had a significant impact on these otherwise upward trends. Tourism across Peru peaked in 2019 at 4.4 million international visitors and has only recovered partially to 2 million visitors in 2022 (see MINCETUR, 2023). Continuing civil unrest now also impacts international travel to Peru, with numbers projected to be below 1 million visitors in 2023 (see Taj et al., 2022).

[3] See Chun and Han (2015); Drozdzewski (2011); and Iglesias (2016, 2017) for language tourism; and see Mascarenhas (2017); McLennan (2014); Simpson (2004); and Wearing and McGehee (2013) for volunteer tourism.

through colleges and universities, language institutes, Latin America interest groups, word-of-mouth, and eventually social media. The first Spanish language school in Cusco was established by a married local/expat couple in 1996, and within 10 years there were half a dozen. In the mid-2000s, Cusco started being billed as an ideal destination to study for TEFL (Teaching English as a Foreign Language) certification due to the city's livability, aesthetic charm in its mix of colonial and Incan architecture, comparable affordability in relation to larger Latin American urban centers, and recreation opportunities.

The growth in the volunteer tourism industry niche came slightly later, supported by the "NGO-ification" (Lewis, 2005; Schuller, 2012; Sinervo & Cheney, 2019) of the social assistance landscape. The late 1990s and early 2000s were marked by the retreat of the government and the establishment of non-governmental organizations (NGOs) catering to a host of needs. These include NGOs involved in classic development projects like healthcare, construction/infrastructure, and education. Interestingly, these NGOs, like Spanish schools, were also founded in many cases by married local/expat couples. Investment in Peru was on the rise globally, as funding from European governments and intergovernmental agencies like UNICEF created pathways for grant money. Peru's reputation as one of Latin America's poorest countries, combined with careful marketing of its cultural heritage touring opportunities, created a compelling portrait for long-term tourist immersion on the volunteering front as well.

Firmly linking language tourism with volunteer tourism came a few years into the process of building Spanish language schools, according to early industry pioneers. One chief goal was to ensure that language tourists were given the opportunity to practice their evolving conversational Spanish. There was also a push to spread out the benefits of tourism to a broader local audience by using foreigner volunteers to support the growing landscape of NGOs who needed both labor and donations. Narrated as a mutually beneficial endeavor, Spanish language competence would create a more meaningful interaction between locals and tourists, as well as ensure more efficient use of volunteer assets. It made sense that for-profit Spanish schools would be the connectors, providing Spanish language classes and home-stays alongside volunteering opportunities. Yet, an unpredicted byproduct of using NGOs as sites for tourists to practice

Spanish conversation is that local Peruvians would also harness opportunities to practice their English.

Economic livelihood opportunities in Cusco are heavily leveraged in the tourism industry, with English language proficiency as a requirement in most formal service sector jobs like restaurants, hotels, travel agencies, stores, and guiding (Hall, 2017). Informal sectors also rely on English, whether selling artisanal products and handicrafts, offering guiding by foot or taxi, or hawking touring and dining opportunities in the streets. The ability to offer a product or service in the "global language" of English (Salazar, 2006) provides a tangible boost in income now and potential earning power later.

Working, Playing, and Learning with Tourists: Cusco's Child Vendors

Children have long been a part of Cusco's informal working sector, often perceived by tourists as mere accompaniments to their laboring parents, who do not have other childcare options and elect to bring their children along while they work. In fact, there is a deep history of children's involvement in household public work in the Andes (Baufumé, 1998; Bolin, 2006), and as rural to urban migration intensified, families brought this work ethic and set of expectations for childhood to the city and the jobs they found there (Campoamor, 2016; Leinaweaver, 2008; Taft, 2019). Children in turn have taken up the opportunities afforded them in the city streets and created their own brand of *ambulante* street vending, exponentially profitable due to the creative and emotive appeal engendered by Western notions of childhood (Sinervo, 2011a; Steel, 2008). Child vendors start off as babies and toddlers carried by their mothers as they sell, and they might begin to make their own sales around the age of four or five. By age seven, many child vendors work with groups of siblings or friends, and quickly they also begin to work alone.

The school day for children in Cusco takes place in five-hour shifts, so there are opportunities for studying, selling, and socializing after or before school. Many NGOs for children in close proximity to the city center are set up as drop-in afterschool programs staffed by volunteer tourists, who were in turn placed there by their Spanish schools. Volunteering shifts can last for weeks or months, and might involve a particular tourist coming and going as they engage in our other travel opportunities around the region.

Since volunteers are tasked with supporting children's academic progress, they often assist with homework help and designing extra-curricular enrichment. At one program, there is a rotating curriculum intended to expand children's worldviews, for example, through learning about world religions. Volunteers often offer, or are asked, to teach English. Yet volunteers also want to "help children to have childhoods," so there is a significant emphasis on play activities such as art, games, sports, and even dance parties. Children and volunteers can connect quickly—even across dramatic linguistic divides—and that connection becomes layered with affectivity and intimacy: tourists are primed to seek authentic engagement and children desire friendships with generous foreigners.

Children create their own playing and learning activities when undertaking their street vending. It is not uncommon to see children playing jacks or drawing with tourists in a corner of the main Plaza, or engaged in elaborate games of hide and seek or tag across the city space. Some children report that they come to sell so that they can socialize with their friends who live and go to school in other parts of the city. Learning happens through vending interactions as well, such as in the opening vignette to this chapter where I emphasized that children treat interaction with a tourist as an opportunity to practice English and expand their global knowledge. In sum, working, playing, and learning happen simultaneously in most encounters between Cusco's child vendors and international tourists. Moreover, children are key actors influencing touristic experiences across cultural, economic, generational, and linguistic divides.

Methods: Accessing Foreigner and Local Points of View

The ethnographic portraits in this chapter are compiled from research conducted in Cusco over 20 months between 2005 and 2017. Participant observation took place in the streets around the main city square (the Plaza de Armas, colloquially termed "the Plaza"), in two afterschool programs located a 5 to 10 minute walk from the Plaza, in three Spanish schools, and at tourist venues like restaurants, bars, and clubs. I also participated in networking meetings for children's rights organizations and visited half a dozen other child-focused NGOs, two of which feature in this chapter.

Interviews were conducted with children, their parents, tourists, Spanish school volunteer coordinators, government officials, and NGO center directors and staff. To further access tourists' points of view, I conducted focus groups at local Spanish schools and restaurants with four to six participants at a time, and gathered approximately 200 multiple choice surveys focused on tourists' casual and vending interactions with children. When working with children, I solicited their writing (journaling and essays), drawings, maps, kinship charts, and photographs that explored various themes related to my research. Children also participated in peer-facilitated interviews, which were recorded, transcribed, and analyzed.

I collected city records listing NGOs and reviewed local and national laws on informal work, city space, and children's rights. Statistics at the local, regional, and national level were accessed through government divisions in Cusco and Lima, as well as from the NGO programs themselves. While gathering these data, I also interviewed a handful of government officials and several pioneers in the volunteer tourism and children's aid landscape.

Teaching and Learning Fluctuations, Linguistic Dreaming, and English as Enrichment

Two months into a year-long fieldwork stint in 2007–2008, I was invited to spend time at a police "family center" for children rounded up in the streets. There I witnessed a rotating host of volunteers, supported by a local NGO afterschool program and placed by three different Spanish schools, attempting to engage the children in custody who ranged in age from 8 to 16 years old. Some children had been there for weeks and others only for a few days or hours. All were waiting to be "claimed" by guardians after being noticed for doing illegal work like informal street vending or for living on the streets. There were rumors that some children were being held[4] for suspected "delinquency" like drug use or theft, but all such accusations were kept quite vague; it was clear that children were awaiting guardians rather than the justice system. The supervising police officers focus primarily on essentials like meals and hygiene, while volunteers were present to make the children's stay "more pleasant." The repertoire for engagement seemed to be games and English lessons, beginning right after breakfast.

The English lesson started when one of the four volunteers announced that any child not wanting to learn was welcome to go to the adjoining dormitory because classes are offered based on "voluntary participation." Somewhat surprisingly to me, 4 of the 18 children departed, although they stood on the beds in the dorm and called out to us in English—trying to engage and banter with volunteers—through the windows connecting the rooms. I was told that these children have been at the center for the longest period of time: volunteers surmised that long-time residents are bored with the same English-learning routine.

The volunteers grumbled among themselves that they wish they could build a progressive language curriculum, but since the children are always coming and going, it is hard to have a consistent set of lessons. A newer volunteer from Texas eagerly proposed developing two different English class tracks: one progressive and one basic, because there seem to be enough volunteers to sustain hosting two separate classes! The volunteer in charge, a no-nonsense Dutch woman, glared stonily at him while pointing out that volunteers themselves change over. She later justified her cold reaction by complaining privately to me that the Spanish school placing him tended to send volunteers who stay for too little time, with too little Spanish proficiency, and too much entitlement and expectation to receive "easy work" because of the high rates they pay for their

[4] Nomenclature matters here: police I interviewed would reiterate that they were not allowed to legally "detain" children, but they could facilitate child-family reunification when children were found to be in conditions of suspected abandonment (such conditions being problematically extrapolated based on children's perceived status as unaccompanied in public space).

volunteering experiences. She claimed that these limitations make volunteers like him "pretty much useless" as planners, teachers, or even workers in the program. "It is like a train station around here," she emphasized, "with both children and volunteers coming and going."

The remaining 14 children sat more or less eagerly in their broken plastic seats, practicing a conversational phrase: "Where are you going? I am going to …" One grinning younger girl said that she is going to the cemetery. Other children offered various locations in Spanish (school, work, hospital, police station, stadium, shop, home) and asked the volunteers to translate. Having run out of momentum on places, the lesson veered toward verbs, as the volunteers pointed out that in English you can be going to a place or going to do an action: "I am going to eat," "I am going to drink," "I am going to knit" (a girl was knitting a jacket in the corner while she listened). You can go to a country too—and suddenly we were back discussing places. This was a favorite topic for the children, who would love to be going anywhere else that doesn't involve being stuck in the police detention center, but would especially love to be visiting the countries that these tourist volunteers come from.

After a rowdy 15 minutes of discussion, where the volunteers worked to entice different children to participate—and the children in the dormitory next-door continued their heckling—the room quieted somewhat for practice writing sentences using the conversational phrase template and drawing pictures about where the children were going or what they were going to do. About half the children were continuing to work enthusiastically. A small cluster of girls sat together at the back, talking and crying—and children continued to come and go as guardians picked them up or new children were brought in by police officers patrolling the city streets. Meanwhile, the remaining children who kept drawing were going to the United States or to Spain; a child drew a war between Spain and Peru. Some were going to their own house, their own bed. One of the oldest children was "going to drink": he drew a donkey in a business suit sitting at a fancy restaurant.

Coming and Going: Fluctuations in Volunteer Teacher and Child Learner

While a police detention center is a particular kind of space, the maneuverings of children and volunteers in the above vignette are replicated at many afterschool programs throughout the city that are operated on a drop-in, free basis. Children come and go, but so do tourists. With these unstable and inconsistent populations of teachers and students, how does one make inroads into learning—or teaching—a language?

Spanish language schools who place volunteers recognize this problem. In interviews with volunteer coordinators, they recounted efforts to enforce formal and supervised language learning, such as requiring two weeks of Spanish language classes *before* being placed as a volunteer, and/or requiring continuous language training as a concurrent complement to volunteering. They might also expect placement stint minimums, such as for two weeks or one month. The assumptions of this model are that increased Spanish proficiency correlates with more effective volunteerism.

Spanish schools narrate the benefits for multiple stakeholders that stem from interactions influenced by formal tourist language instruction—this is similar to processes that Phoebe Everingham highlights in other volunteer tourism scenarios through a framework of "mutuality" prompted by "intercultural learning" opportunities (2015, p. 176). Tourists deepen their language learning and augment their travel experience with increased immersion and relationship-building. Organizations and child beneficiaries are able to depend on volunteer labor that is less fleeting and offers more impressive ongoing and cumulative teaching/learning opportunities (such as the proposed dual track English program at the police station). Finally (and more quietly shared with me by my more jaded interviewees), Spanish schools—for-profit businesses who sell language classes and volunteer placement opportunities as part of their revenue plan—benefit economically.

Nevertheless, it is important to understand the intricate economics of the relationships among Spanish schools and students/volunteers, volunteers and children's aid organizations, and Spanish schools and children's aid organizations (Sinervo, 2015). In juxtaposition with dominant marketing discourses that highlight volunteer tourists' potential for contributions through meaningful labor (see Jakubiak, this volume; as well as Jakubiak, 2017, pp. 206–207), many volunteers are accepted into programs chiefly for their economic resources.

An aid program will welcome a volunteer that they cannot train, supervise, or use so as to receive a fee from the Spanish school, who donates to the program per volunteer placed. Volunteers may be unhappy because they do not have meaningful work—a challenge I address further below—and programs may be unhappy because a single volunteer placed for a longer time generates a *single* donation rather than the *multiple* donations that could be received by placing several shorter-term volunteers.

This funding model creates jealousy rather than partnerships between Spanish schools and aid programs. I was dragged into an ongoing feud, as volunteers at one school and program pairing angrily pointed out they would rather their "overpriced" volunteering fees go "directly to the children" than be filtered and depleted through an intermediary (see McLennan, 2019 for another discussion on volunteer commodification). Despite Spanish school officials' rejoinders on the apparent connections among time depth, linguistic confidence, and volunteer efficiency stemming from concurrent language classes and volunteering, we can clearly see in the "train station" quality of the police family program that volunteer continuity is hard to enact in practice anyway. Longevity is one complex challenge, but it sits alongside the heavier accusation of unfair resource redistribution practices across the volunteer/language tourism pipeline.

One could argue that a volunteer might be perceived as more valuable for their "fee" than for their unskilled teaching labor, yet this would be overlooking the affective and intercultural advantages of the teaching and learning interactions between children and tourists. In this aspect, longer volunteer placements do win against any measure: children and tourists feel more connected when they spend more time together. Paradoxically though, I see relationships cemented in intensity not only through continuity but also through parting and reunification. At one program, a beloved volunteer left and returned multiple times; each time she returned, children spoke of feeling more assured that she "loved" them and would "keep caring" about them from a distance, even when she was not there.

In the police family center, the fluidity of children as language-learners also frustrated volunteer efforts to deliver linear educational content. For the volunteers, a child leaving the program meant a lost chance to enrich their lives through play, affection, and language learning—a missed opportunity to (aspirationally) build consecutive linguistic skills—but for the child waiting for a family member to pick them up, English classes were a way to pass time during a dreary life interruption. I noticed this when a volunteer would tearfully hug a "favorite" departing child who had been cuddling them the hour previously, yet the child was simultaneously trying to reach around the volunteer to embrace a parent or older sibling who had finally arrived to collect them. This disjuncture in program valuation aside, there is also a challenge with getting *all* the children onboard taking English language classes that are "optional"—as we can see in the dichotomy between the 14 "newer" children who playfully engaged the lesson and the 4 long-time, bored children who refused the lesson but still desired conversational interaction. The latter group is therefore ironically also a fluctuating group of learners: while around the longest and thereby able to possibly benefit from a curriculum that offers progressive levels of instruction, they do not *consistently* want to participate. I return to this mediocre enthusiasm for English language classes further below.

English as the Language of Mobility: Linguistic Dreaming of Cusco's Child Vendors

It is not coincidental that the children at the police station use an English lesson as a chance to dream about their physical and cultural mobility. There are two analytical directions that should be teased out here. The first—and likely more obvious—is that local children would love to be able to travel as tourists do. As a population thoroughly enmeshed in the commerce, interactions, and potential relationships of the tourism industry, children in Cusco seek to learn about tourists' lives and languages in order to build cultural capital for that potential future where the children might too travel and immerse themselves in different ways of life. The second insight is that children also use English to be better economic participants in Cusco's tourism sector. Children have dreams but are not naïve. They understand that the pathway to a future where they might one day be tourists themselves requires hard work right now. Speaking English with tourists facilitates making sales, earning a livelihood that helps support the vendors' families. English is not just used to imagine mobility but also to enact it through contemporary street selling practices.

Twenty years ago, Mercedes Niño-Murcia made a similar point eloquently in her text focused on how "English is like the dollar" in Peru. She writes:

> Learning English has become a component of an "imagined global citizenship," one of the many ways of "imagining globalization." ... Because it is seen as the linguistic currency that makes one a potential actor in the global marketplace, English has acquired special—perhaps inflated—standing among forms of cultural capital. ... The perception of the English language as the strongest linguistic currency in today's society, or more pointedly, as an equivalent of the American dollar, as something that needs to be attained in order to participate as a consumer in the global market, has transformed the study of English from an instructional activity, a tool for learning, into an object of consumption. (Niño-Murcia, 2003, pp. 121–122)

Likewise, children in the police family program have a desire *for* English which they express *in* and *through* language learning opportunities. English learning facilitates linguistic dreaming about upward mobility. Nevertheless, Niño-Murcia (2003) also points out that English skills alone are a false promise for escaping poverty. English comes alongside existing wealth and can only be strategically useful for those who are already positioned to use this capital. Moreover, English as a skill does not facilitate *actual* mobility, such as access to tourist visas and immigration processes. Yet youth in Cusco today continue to view English as the pathway to a better job, which will earn them more money and in turn give them access to travel. Speaking English is a part of cultivating oneself too, the "linguistic entrepreneurialism" mentioned in the introduction (De Costa et al., 2021). With this desirability in mind, why did some of the police center participants reject English language classes?

Linguistic Power Struggles: Classes Versus Practice as Forms of Language Enrichment

While compulsory English is taught in public schools in Peru (Niño-Murcia, 2003, p. 127), children desire English learning from tourists, connecting to the "native speaker fallacy" of effective instruction. While this concept is defined in terms of the supposed naturalness of the English *speaker* as the English *teacher*, it also connects to the assumption that just

> knowing English is a necessary *and sufficient* qualification for securing well-paid employment in the "competitive world of work," a place where the ability to engage in waged labor is not a right but must be earned and occurs predominantly in the private sector. In this frame, development no longer means structural changes that could reduce many peoples' vulnerability. Instead, development means helping individuals navigate economically volatile terrain alone by equipping them with the value-added skill, English. (Jakubiak, 2020, pp. 221–222)

Since children would agree that English-speaking tourists are appropriate teachers of English—and that English is the pathway to mobility—at the heart of disputes over English language *classes* is actually dissatisfaction with the mode of learning. Volunteer tourists—untrained, uncredentialed, inexperienced, and transient—see English language teaching predominantly in the mode of their own experiences with "English language arts" in school: a focus on learning grammars and scripts, perhaps geared toward mastery in formal written assessments, attained through progressive curricula. Child vendors, on the other hand, want English *practice* in order to become better conversationalists with higher potential for success as current and future workers in the tourism industry.

The children at the police center who I witnessed refusing English language classes were legible as bored with the same old learning, but the volunteers interpreted this boredom as one based on progressive curriculum limitations: failing to see that as children continued to converse through the dormitory windows, they were seeking relational engagement in oral, usable English. Structured lessons are what tourists think they should teach, but friendly banter is what children want to use for language learning.[5]

These insights on what children are looking for as language-learners are consistent with comments made by the majority of vendors when comparing Peruvian public school English classes with the English learned from tourists; however, the insights make more sense when you *also* consider what children say when comparing volunteers' English classes with opportunities to interact with tourists casually in street interactions. Learning English from foreign tourists is better than learning English from Peruvian teachers in school, but practicing English "enrichment" in the streets is *even better* than receiving English instruction in children's aid programs staffed by volunteer tourists. Practicing English is actually highly ranked as one of child vendors' chief motivations for engaging in *ambulante* street selling: sometimes ranked higher than earning money. Given that this (economic and linguistic) practice might come at the cost of limits to their physical freedom—not least at a place where they are ironically offered less desirable forms of English engagement!—it is worth paying attention to not just the allure of English but rather to what children are really looking for when they consider tourists to be ideal language practice partners.

[5]While beyond the scope of the current analysis, I also viewed children's English class nonparticipation as protest over their involuntary detention at the police program. In some ways, children saw volunteer tourists as their allies (capable of bringing them additional resources and compelling better treatment from the national police), but in other ways, tourists were accomplices in making the program seem like a legitimately nurturing place (for children to be held, separated from their families).

Meaningful Work, Meaningful Learning

Many of my interviews with volunteer tourists echoed the comment made by the Dutch supervisor at the police project: volunteers are paying for a *particular kind of experience* and want their money's worth. There is a tension here between opting to participate in service tourism in order to engage in social justice (Sin et al., 2015) versus to facilitate feelings of one's own ethical goodness through moral consumption (Smith, 2014). In either framework, volunteers are clients, and whether or not they rationalize a high volunteering fee as an investment in a personally rewarding tourism experience, it would be simplistic to claim that all volunteers are looking for "easy" work for their money, since many instead articulate a desire to provide "meaningful" contributions.

Take the experience of a long-term college-aged volunteer from Canada. Arielle was taking a semester off to travel in Latin America after studying abroad in the Netherlands and the UK. She arrived in February as a volunteer in one of the afterschool centers catering to vendors in the city's main Plaza de Armas.[6] On her second day of work, she struggled with her limited Spanish to help a child understand a complicated English grammar concept they had been assigned for homework. Later she worked with another student on practicing English reading: Arielle would read first and then encourage the child to read back to her.

Within the week, Arielle was spending more time designing English lessons for the students. She had decided—and been encouraged by her volunteer coordinator—that this was a way that she could practice her Spanish while also making a meaningful impact on the children's learning. Over drinks before a local pub quiz, whose proceeds would go to support a different afterschool program, Arielle reported that the volunteers staying at her bunkhouse were disillusioned that they were not working hard enough: "It is not rustic at all like we expected. We are living in this nice house where people come to feed us and clean up after us, and we are only volunteering for four hours each day. We are too

[6] While this center is also run by the national police, it is unlike the "family center" at the station because children here are free to come and go on a drop-in basis.

comfortable; we did not expect it to be this easy."[7] Teaching English thereby became a way to "create work for herself": to extend the value of her experience and the potential value she could provide for the local children.

Unfortunately, this was still not enough for Arielle: a week later she had started a second volunteer placement at an orphanage outside of the city so she could "use her time meaningfully." The leveling up did not go as planned as she complained another week later of having been bussed to a site two hours away where she had been asked to teach three different English classes with no advance preparation. "I did *not* like that experience," she reiterated at the next Tuesday's pub quiz. "The children were hectic and out of control, and I have asked for a transfer to somewhere closer to spend less time commuting and more time teaching."

Arielle's next project was more fulfilling; a month into her placement, she brought me with her to visit and I had the opportunity to watch her run one of her English lessons. The children enthusiastically collected notebooks and pens, sitting down seriously to listen. Arielle began by reviewing material she had previously taught, which included colors, numbers, asking simple questions, and introductions using names, ages, and birthdays. She transitioned to focusing on how to conjugate "to be." The older children were engaged and asked questions, while the younger children mostly drew as they listened.

After a 30-minute lesson, Arielle wrapped up and suggested that we touch base with the newer volunteers. She would be leaving Peru within the week and had heard from the children that she had been their most consistent English teacher that year. The children and Arielle were fearful that a new English teacher would start them back on learning numbers and colors again, not supporting them to progress further in the language. Arielle was hopeful that orienting the new volunteers to what she had taught would allow them to build on her contributions.

[7] For another ethnographic example of how placement sites can be discredited by volunteers as "too nice," see Jakubiak (this volume).

Volunteers as Experience Consumers: Entitlement to Meaningful Work

A rich literature on volunteer tourism struggles with the "leisure-volunteer duality," whereby tourists expect entertainment as well as real-life opportunities to help others and the role this expectation plays in "sensemaking" about their experiences (Liston-Heyes & Daley, 2017). Tourist satisfaction with their experiences can relate directly to a costs and benefits analysis (Hallmann & Zehrer, 2016) of how one simultaneously helps the self while also helping the other. Erica Bornstein writes of volunteer tourism as an experience that "exceeds the realm of routine everyday requirements ... enter[ing] into an arena of free action that is uncoerced and selfless [and thereby] urges toward expression or communication with others ... [in] situations that are formative" (Bornstein, 2012, p. 119). Yet she also reiterates that while it is "an inspirational model for social engagement ... it is not an easy endeavor" (Bornstein, 2012, pp. 120–121). Finally, Bornstein profiles how tourists are "encouraged to focus on the experience of the moment, not the historical or political economy that created the moment" (2012, p. 137).

Arielle's story illustrates that tourists seek to add value in many different ways, but they often also crave structure and measurable impacts. This is particularly the case in contexts that feel highly unstructured and where they are not sure how to help. When directly asked what skills they bring to an organization, tourists might be primed to offer their English-*speaking* abilities because this is the messaging that they receive about their assets from their Spanish school volunteer coordinators as well as from the local staff at the organizations. When volunteers are eagerly seeking a clearly defined way to contribute, it becomes easy to mistake the possession of native English with a (potentially non-existent) expertise as a language teacher.

I hypothesize that these erroneous assumptions are partially a result of staff not knowing what to do with the volunteers they solicit. As I mentioned earlier, volunteers may be more desirable for their fees than their work. However, the assumption of expertise spotlights that another significant problem is tourists' lack of critical reflection on how English language volunteer tourism reproduces colonial dynamics, legacies of inequality, and othering, especially in the positioning of English as a world language that symbolizes dominance and hierarchical differences (Thomas-Maude et al., 2021; see also Garlan, 2012). Tourists understand that they possess something valuable (English) that they are structurally positioned (as adults) to share, which blinds them to the plausible limits of their assets (as good teachers). Tourists also do not want to be told that their money is more beneficial than their labor, since that delegitimizes the volunteering component of their travel experience.

We are further left wondering about how Arielle links meaning and challenge. Can work be meaningful if it is not challenging? There are clearly gradations in this, as Arielle found one work site to be *too challenging*: a situation she characterized as children "out of control," which we might ungenerously index as stemming from her inexperience with classroom management rather than from the children's behaviors in a perhaps unstimulating and inappropriately leveled learning environment. Jacob Henry might label Arielle as an "underprepared teacher," similar to a "maverick [...] white savior" teacher that shows up in Hollywood films and inadvertently facilitates "educational violence" (2020, p. 1, p. 8). Such teachers do not have training in how to build coherent instructional goals and methods, or how to facilitate classroom community; they therefore "must run their classrooms primarily on their intuitions about what 'good teaching' looks like," taken from pop culture (Henry, 2020, p. 1). Key features of underprepared teachers include their supposed charisma, their interest in students as specific learners, and their disillusionment with the existing system of education that compels them to create their own unique curricula (Henry, 2020).

While the meaning of the work is defined in how it makes the volunteer feel valuable—even in the face of clear feedback from the recipients about what makes learning meaningful for them—it also intersects with specific linguistic benefits for the teacher. Interestingly, Arielle comes to English teaching after being encouraged by her volunteer coordinator to use it to practice her *own* Spanish. I noticed this too in the interactions between the children and Arielle: everything conversational was done in Spanish, including instructions and explanations to support the lesson, even as the session itself was designed for students to engage in English.

Arielle used her earlier experience—of her *own* struggle with using Spanish to teach English—to segment the languages. She would teach English, but would practice Spanish through the opportunity to connect as a longer-term teacher for a specific student population. In the final section of this chapter, we will turn to the framework of children and tourists as conversation partners—perhaps a dynamic better-suited to their capacities and goals—but first I want to advance a short analysis on why we should go beyond what "meaningful" looks like for volunteers. I suggest that it could be transformational for us to pay attention to what children might be looking for as English language learners. To do this, we must go back to critiquing the supposed structures of power that operate within the children's aid context.

Child Vendors as Agentive Language Users: Desires for Meaningful Learning

When expressing their concern to Arielle that they would have to start all over again when receiving a new teacher, program children seem to echo the critiques of the limited temporalities of volunteer tourism that we have explored above. Repetitive learning structures mimicking basic school curricula emphasize learning colors, numbers, simple questions, and useful verbs such as "to be" and "to have." Since volunteers are defining teaching very narrowly, without critical reflection on what language education is and can be, their instruction is a rote performance rather than focused on what can best achieve student learning. How much children are learning is never measured and tourists are not held to account for whether they are efficient teachers, as one would find in a traditional educational setting.

Yet children are not passive recipients of these problematic language-as-aid enactments. In petitioning Arielle to onboard incoming volunteers, they are helping structure her awareness of the challenges they face based on her transience while also possibly influencing the plans of the forthcoming new teachers. We must realize that when children reject formal language instruction, it is not only based on the kind of learning they find meaningful: they are additionally considering the kinds of interactions that nurture feelings of kinship between child and volunteer. I saw this entanglement of practical, relational, and affective goals when children appealed to volunteers to sit with the children while they learned rather than teach from the front of the room. Children engaged in flattery of volunteers' Spanish-speaking skills, a further method for building rapport and compelling multilingual exchanges in Spanish and English.

Children expressed their desires for meaningful learning when they boycotted or acted disruptively in English classes. In one afterschool program I worked at, certain children would leave the classroom and come sit with me on the outside steps instead, practicing expressions they had heard volunteers using among themselves. They would ask me to translate (English to Spanish) and thereby make legible the words they had heard that day, or to supply them with correct pronunciations for (English) words they were working on integrating into their vocabularies. When I asked (in alternating Spanish or English) why they did not go back inside to participate in the English class, children often said that they felt more comfortable practicing language with me than with the latest short-term tourist who was trying to teach them—they knew me better and I was more aware of their learning needs.

To summarize, I believe that meaningful learning for children avoids rote and unimaginative thematic repetition, facilitates linguistic skills that are immediately applicable to their daily lives, and harnesses modalities that also facilitate closeness between children and tourists (see Doerr, this volume, on neo-immersion). Children want to feel invested in their teachers and want to know their teachers are invested in them: not just as teachers but as people. It is to this imperative of connection that I turn in the concluding sections of this chapter.

Connection Through and In Spite of Linguistic Barriers

Fifteen-year-old Adrian was always goofing off when attending the afterschool program. He was not attending school when I met him, preferring to spend his time working as a musician playing the *zampoña* for tips in restaurants. Adrian had left home due to familial conflict and commuting distance, and rotated sleeping on a mattress in an internet café closet and at a friend's house. My first extended conversation with Adrian was a few weeks into knowing him. He had spent those initial weeks of our acquaintance teasing me using Spanish slang that I did not understand, but our interaction that particular day involved him reading Spanish short stories from a workbook and inviting me to take turns reading with him. He carefully explained the plots of each story after we finished, and I wondered whether he was concerned that I did not understand them or was eager to demonstrate what he had himself learned. Protective of his privacy and suspicious of tourists coming and going, it was four more months before he showed interest in being interviewed. He shared that he had left school initially due to an extended illness and then had not returned because he feared reprisals from having broken a school window playing soccer. While Adrian felt like he had learned plenty before leaving school, he mostly missed the English classes that supported his interactions with tourists and therefore his livelihood.

Six weeks after this interview, Adrian met Sophia, an older American tourist with whom he formed a deep connection when she frequented his restaurants to watch him perform and treat him to meals. Although they struggled with the language barrier, they worked to keep in touch after she returned to the United States following her short two-week volunteer placement. At first, they each wrote emails in the other's language: Sophia spending 30–60 minutes per day studying Spanish and just as much time composing Spanish messages to Adrian, and Adrian writing one to two lines in English that Sophia found difficult to understand. They both used me as an intermediary, Adrian asking for news of Sophia (who emailed me in English) and Sophia asking for help interpreting Adrian's messages which felt cryptic and stilted. After a few weeks, I suggested that they each write in their own language and do their "practice" by reading rather than writing. Sophia thanked me, finding that Adrian was much more engaged when writing to her in his native Spanish.

With a growing bond with Adrian and feeling a lingering affection for many of the children whom she had met, Sophia decided to return to Cusco two months later. She wanted to fundraise for the afterschool program and asked her Spanish school volunteer coordinator intermediaries to find out the NGO's greatest needs. Among them were running water, a larger stock of school supplies, and a permanent English class, which they believed would attract more children to the program and keep them attending more consistently. Sophia's return visit was short like her first, but provided further opportunity to cement her friendship with Adrian. When asking Adrian afterward what he had enjoyed about her visit, he commented on the opportunity to practice his English and help her with her Spanish, as well as watching movies (in both languages) and playing chess at her rented apartment.

As the next months passed, Sophia grew to become a central support in Adrian's life. She sent him money for food and to rent a small apartment close to town. Eventually, she sponsored him taking English classes at a local organization that offers TEFL teacher trainings—so Adrian had access to semi-professional teachers. When Sophia spoke with me about her economic support of Adrian, she struggled with whether she should focus on contributions that enabled eventual self-sufficiency—such as learning English—or whether she should focus on immediate needs like food and housing. Her uncertainty relied on characteristic discourses about the use of direct charity and sponsorship to cope with poverty, employing classically neoliberal tropes like "deservingness," "improvement," and "entrepreneurship." She wanted him to keep working and not to rely on her. Sophia's "gifts" to Adrian eventually rested more and more on language, and with it the promise of "self-betterment." For his part, Adrian leaned into these constructions and deployed them strategically to expand Sophia's feelings of investment, affection, and empowerment to change the course of his life. This was also encouraged by the Spanish school volunteer coordinators who helped to facilitate Adrian's English classes on behalf of Sophia.

Multilingual Resources for Connection

Like the children at the police center, Adrian desires English and connects it to his cosmopolitan future-building project of self-development. He knows that he is falling behind in his English skills as he no longer attends public school, but he supports his own economic and linguistic development using interactions with tourists when he performs his music. As an entertainer, he is aware of the power of representations of youthful need; yet, he is distrustful of fleeting interactions and reluctant to develop intimate ties that might later disappoint. We see this when he keeps me at a distance, using my lack of proficiency in Spanish slang to ridicule me when we first meet—yet we also see that he wants to be a teacher and facilitate my Spanish skills, even while using me to help him catch up in "school skills" he is feeling behind on. With Sophia, he connects quickly because she is interested in his music and also readily provides him with direct material support (buying meals), which is an indication that she is uniquely invested in him. I also believe that my quick friendship with Sophia combined with my hard-earned trust with Adrian allowed him to let his guard down sooner with a second patiently interested and attentive foreigner.

Nevertheless, Adrian's English skills and Sophia's Spanish skills could not keep up with the depth of engagement and detail—let alone the genuine caring—they each wanted to express to the other. Adrian's and Sophia's shared desire to connect despite language barriers echoes Phoebe Everingham's discussion of the embodied borderlands of interaction and bonding between volunteers and children in Ecuador: "Empathy happens in the liminal space of the encounter … [and] includes affective, cognitive and reflexive elements in language exchanges as something new is created through the agency of those involved" (Everingham, 2018, p. 71). When children become "experts" in the teaching of Spanish to volunteers, not only do they reverse the classic power dynamics between tourist/adult/English-speaker and local/child/Spanish-speaker, but they also reframe the value of Spanish itself as a sought-after commodity (see Motobayashi, this volume, on the commodification of Spanish). Both volunteers and children have linguistic resources and they are capable of forming a relationship based on each other's assets rather than deficits. They are able to be both teacher and learner—together—and navigate the cross-cultural misunderstandings that are prone to arise in the typical volunteer tourism encounter (Raymond & Hall, 2008). This progresses through hard work and accommodation—both Adrian and Sophia engage in more language classes to support learning each other's language—and through time spent together.

Meanwhile, they must still navigate well-intentioned intermediaries such as volunteer coordinators,[8] who also profit from Sophia's trust in them to steward her economic sponsorship of Adrian. Furthermore, they have to contend with each other's (mis)conceptions about the other's needs, and confront the opportunities and limits of their relationship based on their own cultural biases about economic and affective giving. Like all the other children I met during my years conducting research on child vendors' interactions with tourists, Adrian deploys his language assets and positionality strategically: he expresses care for Sophia's growth and emotions by helping her practice her Spanish and spending time with her, while also safeguarding his own aspirations.

Linguistic Cosmopolitanisms in Action: Global Languages and Their Values

"English as Peruvians sense it is not some generic world 'English' but English as evaluated within a local language mentality" whereby it has higher cultural capital than Spanish (Niño-Murcia, 2003, p. 126). English is necessary for child vendors to be successful cosmopolitans and economic agents, through connections with and selling to tourists. In the desire for English, we see children dreaming about different kinds of mobilities. However, child vendors know that tourists elect to engage in volunteer tourism as a companion to Spanish language practice, and they also leverage their assets as Spanish conversation partners. Children want to learn through oral practice, not instructional classes: they feel that their English improves faster, but they also find more opportunities to connect. Connection is important for socioemotional *and* economic goals: as we can see in the case of Sophia, tourists are more likely to sponsor a child when they feel kinship. For their part, volunteers want to do meaningful work, but they often misjudge their own skills as language teachers. The power dynamics of afterschool programs might enable tourists to control how they offer language, but in street encounters and

[8] I might add "and anthropologists," who also can profit from careful analytical attention to a particular tourist-child relationship when both parties are willing to contribute to the research project.

resistance to formal training, children restructure the ways that learning and teaching can take place.

In the contestations we have charted among teaching and learning, instruction and practice, and long- and short-term volunteer tourism, we see that both Spanish and English are valued, commodified, and generative of economic opportunity alongside linguistic cosmopolitanisms and affective connections. Spanish and English are "global languages" in negotiation with each other—and after all, teaching language is learning it, too.

References

Baufumé, R. I. (1998). Trabajo infantil: Niños, niñas, y adolescentes construyendo su espacio social. .

Bolin, I. (2006). *Growing up in a culture of respect: Child rearing in Highland Peru*. University of Texas Press.. https://doi.org/10.7560/709829

Bornstein, E. (2012). Volunteer experience. In C. Bender & A. Taves (Eds.), *What matters? Ethnographies of value in a not so secular age* (pp. 119–143). Columbia University Press. https://doi.org/10.7312/columbia/9780231156851.003.0005

Butcher, J., & Smith, P. (2015). *Volunteer tourism: The lifestyle politics of international development*. Routledge. https://doi.org/10.4324/9781315796376

Campoamor, L. (2016). "Who are you calling exploitative?" Defensive motherhood, child labor, and urban poverty in Lima, Peru. *Journal of Latin American and Caribbean Anthropology, 21*(1), 151–172.

Chun, J. J., & Han, J. J. (2015). Language travels and global aspirations of Korean youth. *positions: asia critique, 23*(3), 565–593. https://doi.org/10.1215/10679847-3125913

Crystal, D. (2012). *English as a global language* (2nd ed.). Cambridge University Press. https://doi.org/10.1017/CBO9781139196970

De Costa, P. I., Park, J. S.-Y., & Wee, L. (2021). Why linguistic entrepreneurship? *Multilingua, 40*(2), 139–153. https://doi.org/10.1515/multi-2020-0037

Drozdzewski, D. (2011). Language tourism in Poland. *Tourism Geographies, 13*(2), 165–186. https://doi.org/10.1080/14616688.2011.569569

Everingham, P. (2015). Intercultural exchange and mutuality in volunteer tourism: The case of intercambio in Ecuador. *Tourist Studies, 15*(2), 175–190. https://doi.org/10.1177/1468797614563435

Everingham, P. (2018). Speaking Spanglish: Embodying linguistic (b)orderlands in volunteer tourism. *Emotion, Space, and Society, 27*, 68–74. https://doi.org/10.1016/j.emospa.2018.04.001

Garlan, E. (2012). How should anthropologists be thinking about volunteer tourism? *Practicing Anthropology, 34*(3), 5–9. https://doi.org/10.17730/praa.34.3.694215q342360436

Hall, A. C. (2017). *Framing a lost city: Science, photography, and the making of Machu Picchu*. University of Texas Press.

Hall, A. C. (2019). Savoring nostalgia: Food and the past in Peru's La Casa de Don Cucho. *Anthropology of Food, 14*, 1–20. https://doi.org/10.4000/aof.10106

Hallmann, K., & Zehrer, A. (2016). How do perceived benefits and costs predict volunteers' satisfaction. *Voluntas: International Journal of Voluntary and Nonprofit Organizations, 27*(2), 746–767. https://www.jstor.org/stable/43923201

Heller, M. (2010). The commodification of language. *Annual Review of Anthropology, 39*, 101–114. https://doi.org/10.1146/annurev.anthro.012809.104951

Henry, J. (2020). The cinematic pedagogies of underprepared teachers. *Teaching & Teacher Education, 89*, 1–9. https://doi.org/10.1016/j.tate.2019.102990

Hill, M. D. (2007). Contesting patrimony: Cusco's mystical tourist industry and the politics of *incanismo*. *Ethnos, 72*(4), 433–460. https://doi.org/10.1080/00141840701768276

Huberman, J. (2012). *Ambivalent encounters: Childhood, tourism, and social change in Banaras, India*. Rutgers University Press. https://doi.org/10.2307/j.ctt5hjcq1

Iglesias, M. (2016). The language tourism market system: Conceptualising language tourism. *International Journal of Scientific Management and Tourism, 2*(1), 25–40.

Iglesias, M. (2017). Language travel supply: Language tourism product composition. *International Online Journal of Education & Teaching, 4*(1), 1–17. http://iojet.org/index.php/IOJET/article/view/135/145

Jakubiak, C. (2017). One smile, one hug: Romanticizing "making a difference" to oneself and others through English-language voluntourism. In N. M. Doerr

& H. D. Taïeb (Eds.), *The romance of crossing borders: Studying and volunteering abroad* (pp. 195–219). Berghahn Books.
Jakubiak, C. (2020). "English is out there—You have to get with the program": Linguistic instrumentalization, global citizenship education, and English-language voluntourism. *Anthropology & Education Quarterly, 51*(2), 212–232. https://doi.org/10.1111/aeq.12332
Leinaweaver, J. B. (2008). *The circulation of children: Kinship, mobility, and morality in Ayacucho*. Duke University Press.
Lewis, D. (2005). Individuals, organisations and public action: Trajectories of the 'non-governmental' in development studies. In U. Kothari (Ed.), *A radical history of development studies: Individuals, institutions, and ideologies* (pp. 200–220). Zed Books.
Liston-Heyes, C., & Daley, C. (2017). Voluntourism, sensemaking, and the leisure-volunteer duality. *Tourist Studies, 17*(3), 283–305. https://doi.org/10.1177/1468797616665769
Mascarenhas, M. (2017). *New humanitarianism and the crisis of charity: Good intentions on the road to help*. Indiana University Press.
McLennan, S. J. (2014). Networks for development: Volunteer tourism, information and communications technology, and the paradoxes of alternative development. *PoLAR: Political and Legal Anthropology Review, 37*(1), 48–68. https://doi.org/10.1111/plar.12050
McLennan, S. J. (2019). Global encounters: Voluntourism, development and global citizenship in Fiji. *The Geographical Journal, 185*(3), 338–351. https://doi.org/10.1111/geoj.12318
MINCETUR. (2023). Flujo de turistas internacionales e ingreso de divisas por turismo receptivo. Sistema de Información Estadística de Turismo. Retrieved April 20, 2023, from http://datosturismo.mincetur.gob.pe/appdatosturismo/content1.html
Mostafanezhad, M. (2014). *Volunteer tourism: Popular humanitarianism in neoliberal times*. Routledge. https://doi.org/10.4324/9781315547886
Niño-Murcia, M. (2003). "English is like the dollar": Hard currency ideology and the status of English in Peru. *World Englishes, 22*(2), 121–141. https://doi.org/10.1111/1467-971X.00283
O'Rourke, B., & DePalma, R. (2017). Language-learning holidays: What motivates people to learn a minority language? *International Journal of Multilingualism, 14*(4), 1–18. https://doi.org/10.1080/14790718.2016.1184667

Raymond, E. M., & Hall, C. M. (2008). The development of cross-cultural (mis)understanding through volunteer tourism. *Journal of Sustainable Tourism, 16*(5), 530–543. https://doi.org/10.1080/09669580802159610

Salazar, N. B. (2006). Touristifying Tanzania: Local guides, global discourse. *Annals of Tourism Research, 33*(3), 833–852. https://ssrn.com/abstract=2104229

Schedel, L. S. (2023). Exploiter le plurilinguisme pour vendre une expérience 'monolingue': Le cas du tourisme linguistique à Malte. *Langage et Societe, 178*(1), 75–94. https://doi.org/10.3917/ls.178.0066

Schuller, M. (2012). *Killing with kindness: Haiti, international aid, and NGOs*. Rutgers University Press.

Silverman, H. (2002). Touring ancient times: The present and presented past in contemporary Peru. *American Anthropologist, 104*(3), 881–902. https://doi.org/10.1525/aa.2002.104.3.881

Simpson, K. (2004). 'Doing development': The gap year, volunteer-tourists and a popular practice of development. *Journal of International Development, 16*(5), 681–692. https://doi.org/10.1002/jid.1120

Sin, H. L., Oakes, T., & Mostafanezhad, M. (2015). Traveling for a cause: Critical examinations of volunteer tourism and social justice. *Tourist Studies, 15*(2), 119–131. https://doi.org/10.1177/1468797614563380

Sinervo, A. (2011a). *Appeals of childhood: Child vendors, volunteer tourists, and visions of aid in Cusco, Peru*. [PhD dissertation, University of California, Santa Cruz]. https://www.proquest.com/openview/2a970201f8d10b5e83f789cbeeb0c4c4/1?pq-origsite=gscholar&cbl=18750

Sinervo, A. (2011b). Connection and disillusion: The moral economy of volunteer tourism in Cusco, Peru. *Childhoods Today, 5*(2), 1–23.

Sinervo, A. (2015). Brokering aid through tourism: The contradictory roles of volunteer coordinators in Cusco, Peru. *Tourist Studies, 15*(2), 156–174. https://doi.org/10.1177/1468797614563434

Sinervo, A. (2017). Interdependent industries and ethical dilemmas: NGOs and volunteer tourism in Cusco, Peru. In A. Lashaw, C. Vannier, & S. Sampson (Eds.), *Cultures of doing good: Anthropologists and NGOs* (pp. 142–162). The University of Alabama Press.

Sinervo, A., & Cheney, K. (2019). NGO economies of affect: Humanitarianism and childhood in contemporary historical perspective. In K. Cheney & A. Sinervo (Eds.), *Disadvantaged childhoods and humanitarian intervention: Processes of affective commodification and objectification* (pp. 1–35). Palgrave Macmillan. https://doi.org/10.1007/978-3-030-01623-4_1

Smith, P. (2014). International volunteer tourism as (de)commodified moral consumption. In M. Mostafanezhad & K. Hannam (Eds.), *Moral encounters in tourism: Current developments in the geographies of leisure and tourism* (pp. 31–46). Ashgate.

Stainton, H. (2018). TEFL tourism: The tourist who teaches. *Tourism Geographies, 20*(1), 127–143. https://doi.org/10.1080/14616688.2017.1298151

Stanley, P. (2016). *A critical auto/ethnography of learning Spanish: Intercultural competence on the gringo trail?* Routledge. https://doi.org/10.4324/9781315708232

Stanley, P. (Ed.). (2020). *Critical ethnography and intercultural learning: Emerging voices*. Routledge. https://doi.org/10.4324/9780429280016

Steel, G. (2008). *Vulnerable careers: Tourism and livelihood dynamics among street vendors in Cusco*. Rozenberg Publishers.

Stern, S. J. (1998). *Shining and other paths: War and society in Peru, 1980-1995*. Duke University Press.

Taft, J. K. (2019). *The kids are in charge: Activism and power in Peru's movement of working children*. New York University Press. https://doi.org/10.26522/ssj.v2020i14.2275

Taj, M., Glatsky, G., & Turkewitz, J. (2022). As Peru's unrest chases away visitors, many in tourism fear for their livelihood. In *The New York Times*. https://www.nytimes.com/2022/12/19/world/americas/peru-cusco-unrest-tourism.html

Thomas-Maude, J., McLennan, S., & Walters, V. (2021). Cultural exchange during English-language voluntourism (EVT) in Lima, Peru: A postcolonial analysis. *Journal of Language, Identity & Education*. https://doi.org/10.1080/15348458.2021.1991802

UNESCO (2023). *City of Cuzco*. Retrieved April 20, 2023, from https://whc.unesco.org/en/list/273

Wearing, S. (2001). *Volunteer tourism: Experiences that make a difference*. CABI.

Wearing, S., & McGehee, N. G. (2013). Volunteer tourism: A review. *Tourism Management, 38*, 120–130. https://doi.org/10.1016/j.tourman.2013.03.002

5

The Off-Duty Expectations of International Volunteer Language Teachers: A Middling Transnational Perspective

Kyoko Motobayashi

Introduction

This chapter reports on part of a larger study that examines the mobility and trajectories of international volunteer Japanese language teachers (JLTs) who were dispatched from Japan to South American heritage Japanese communities through a state-sponsored international volunteer program. The participants in the present study are a group of individuals based in Japan who—at the time of data collection—were moving to South American countries for two years as volunteer JLTs. In making their *choices* to be volunteer teachers under the particular conditions in which they were situated, the JLTs demonstrated personal agency by identifying themselves as native speakers of Japanese and making use of their native speaker status as a resource for achieving their transnational mobility. Their choices regarding transnational mobility were closely

K. Motobayashi (✉)
University of Tokyo, Tokyo, Japan
e-mail: kmotobayashi@g.ecc.u-tokyo.ac.jp

L. S. Schedel, C. Jakubiak (eds.), *Voluntourism and Language Learning/Teaching*, Palgrave Advances in Language and Linguistics,
https://doi.org/10.1007/978-3-031-40813-7_5

linked to their longer-term professional and personal goals. Upon their encounter with state discourses on Japanese emigrants and their descendants, they also adopted identities as JLTs for *heritage learners* (Motobayashi, 2015, 2016, 2022). These previously published studies have forged an image of a rather linear and singular trajectory of the younger-career, volunteer JLTs.

Yet, there is an aspect missing in these other studies: volunteer teachers' expectations of life outside of teaching duties at their destinations. Volunteer language teaching work was the official and primary rationale for my study participants' transnational mobility. However, as shown in this chapter, this official rationale was combined with expectations of how volunteers could spend their "leisure" (i.e., off-duty) time at their destinations outside of Japanese language teaching work. By focusing on this aspect, which is narrated in participants' interviews as "additional" to their main jobs in the broader experiential framework of volunteer language teaching, this study complements the abovementioned analyses. Through an analysis of the volunteer participants' interests in possible off-duty gains, the chapter aims to shed light on volunteer JLTs' transnational experiences in a more holistic way, showing how expectations for life outside of work form part of the complexity of the choices of the research participants, who attempted to situate their particular volunteer programs within personal mobile trajectories.

The guiding questions of this study are as follows: What are volunteer JLTs' expectations for life outside of work at their destinations? In what ways are volunteer JLTs' expectations related to their decisions regarding language and mobility? As explained below, these questions fit well into the discussion of middling transnationals (Clarke, 2005; Conradson & Latham, 2005a; Ho, 2011; Raj, 2000; Thompson & Tambyah, 1999) and middling transnationals' work-leisure configurations at a transnational scale in the form of "travelling workers" (Uriely, 2005) and "working tourists" (Jarvis & Peel, 2013). This line of discussion has been explored and developed in mobility studies. However, language issues are understudied in the discussion on middling transnationals. Thus, the linguistic aspect of volunteer JLTs' off-duty expectations will be explored in this chapter.

Middling Transnationals and Work-Leisure Configuration: An Emerging Practice of Combining Leisure, Labor, and Mobility

The idea of "middling" forms of transnationalism and related practices of work-leisure configuration began to be explored in the 1990s (Conradson & Latham, 2005a; Raj, 2000). The 1990s was when traditional dichotomies surrounding mobility and tourism, such as work/leisure or elites/labor migrants, began to be destabilized—particularly in the context of postmodern tourism. Prior to that, transnational mobility was largely understood dichotomously, in terms either of the mobility of transnational elites and developing-world migrants (Conradson & Latham, 2005a, p. 229), privileged cosmopolitan elites and translocal semiproletarians, or tourists and vagabonds (Bauman, 1998; see also Juffermans & Tavares, 2014, p. 6). After the 1990s, research trends have gradually shifted to focus on the "middling-ness" of some transnationals beyond such dichotomies (Conradson & Latham, 2005a).

Middling transnationals are defined as expatriates who "belong to the middling spectrum of transnational migration" (Ho, 2011, p. 118) and who are in the "messy, middle range of possessed cultural capital" (Thompson & Tambyah, 1999, p. 217). They are in the "upwardly mobile, middle quintiles" and "more prone to experiential tensions and conflicts than ... high cultural capital consumers [observed in previous research]" (Thompson & Tambyah, 1999, p. 217). In many cases, middling transnationals are "middling" in terms of socio-economic and class positions in their countries of origin, which is reflected in their educational and occupational standing in their home countries. In the societies to which middling transnationals travel, they tend to be "very much of the middle" as well (Conradson & Latham, 2005a, p. 229, see also Collins, 2014). As David Conradson and Alan Latham lamented, until the mid-2000s, this form of mobility was "relatively unexamined" and "surprisingly little is known about these kinds of migrants ... [although] they too are part of the transnational ecumene" (2005a, p. 229).

A notable characteristic of the "middling" transnational mobility is a configuration of both economic and experiential imperatives. Unlike traditional labor migrants, the migration of middling transnationals to new destinations "cannot be understood within a straightforward economic rubric" (Conradson & Latham, 2005b, p. 288, see also Collins, 2014). Rather, "non-economic factors that organise their particular patterns of global movement" emerged as a key component to understanding their mobility (Conradson & Latham, 2005a, p. 230; see also Conradson & Latham, 2005b). Hence, middling transnational mobility tends to be a more nuanced combination of economic and experiential interests as well as working and traveling practices, as mobile subjects actively seek both experiential and economic opportunities transnationally (Conradson & Latham, 2005b).

The work-leisure configuration practices by transnationals, which blur the boundary between work and leisure in destination sites, have attracted researchers' attention since the 1990s. In early 1990s, Ian Munt (1994) analyzed the then-emergent tourism practices among a new middle class and observed "[t]he commencement of professionalization processes in consumption" in tourism—that is, a process in which "the separation of occupational professionalism and consumption and leisure [was] beginning to blur" (Munt, 1994, p. 112). With the increase in tourist-minded individuals seeking unique experiences, who, at the same time, needed a means to support themselves and reasons to justify their travelling, researchers coined the terms "travelling workers" (Uriely, 2005) or "working tourists" (Jarvis & Peel, 2013) to focus on this phenomenon. These concepts emerged together within broader discussions of "de-differentiating" or "de-differentiation" tourist experiences and everyday experiences (Uriely, 2005, p. 203; see also Clifford, 1992), travellers' "dwelling within travel" (Clifford, 1992; see also Conradson & Latham, 2005b, p. 300; also cited in Clarke, 2005, pp. 309–310), and the "intellectualization" and "professionalization" of tourism (Munt, 1994, pp. 110–114), which unfolded in the academic context of what Natan Uriely (2005, p. 203), Ian Munt (1994, p. 101) and Scott Lash and John Urry (1994) call the "postmodern tourism" approach. This research together indicated certain changes in peoples' mobility experiences in contemporary society. Still, as Jeff Jarvis and Victoria Peel (2013) argue, the existence of working tourists continues to "challeng[e] the traditional temporal and spatial segregation of labour and leisure in many developed economies" (p. 114). In that sense, the working tourist phenomenon has continued to capture important aspects of global transnationalism.

Concrete examples discussed under these categories of transnationals include participants in working holiday programs (Clarke, 2004a, 2004b, 2005; Haverig & Roberts, 2011; Jarvis & Peel, 2013), as well as workers in other kinds of mobility schemes prepared by tourism and service industries (Duncan, 2008; Wilson et al., 2010). Empirical studies include young European budget travellers at Canadian ski resorts (Duncan, 2008), working tourists from New Zealand living in the UK (Wilson et al., 2010), young New Zealanders in London (Conradson & Latham, 2005b), EU graduates in Manchester (Kennedy, 2010), or Western youths teaching English in Korea (Collins, 2014). This body of work illustrates the ways in which economic and experiential imperatives as well as work and leisure practices are combining on a transnational scale in the experiences of individual travelling workers or working tourists. These studies show how distinctions are now blurred between tourism and everyday experiences, between travelling and dwelling (Clarke, 2005; Conradson & Latham, 2005b; see also Clifford 1992), and between occupational professionalism and the consumption of tourism (Conradson & Latham, 2005b; see also Lash & Urry, 1994; Munt, 1994; Uriely, 2001, 2005).

The detailed analyses of "transnational lives of the middle" (Clarke, 2005, p. 307) have further elaborated on postmodern tourism scholars' work (such as that by Clifford, Munt, and Uriely), not merely by providing cases but by shedding light on additional aspects such as uneven mobility (Clarke, 2005), the impacts of international tourism and service industries (Duncan, 2008), and the centrality of cultural aspects for working tourists—both at the levels of national and individual identity (Wilson et al., 2010). Other studies illustrate how participants view overseas experiences as central to reaching adulthood (Conradson & Latham, 2005b), as well as the tensions that mobile subjects experience between discourses of exploration and self-development through international mobility on the one hand and constraints posed by their actual socio-economic circumstances on the other (Collins, 2014). These studies have deepened our understanding of the practices of middling transnationals, who tend to be in the "messy middle" (Thompson & Tambyah, 1999, p. 217) in mobility studies.

Volunteer Tourism and Working Tourism

Conceptual differences and overlaps between volunteer tourism and working tourism also need to be mentioned here. The concept of volunteer tourism was forged as international volunteer work expanded as a platform for work-travel around the world (Tomazos & Butler, 2009; Tomazos & Cooper, 2012). Initially emerging as "an individual altruistic endeavour," volunteer tourism has developed and transformed commercially to form a rapidly expanding market (Tomazos & Butler, 2009, p. 196). As Konstantinos Tomazos and William Cooper (2012) have pointed out, ambiguity remains regarding the definition of volunteer tourism. The broadest definition, provided by Tomazos and Cooper (2012) upon a review of definitions in the body of literature, is "volunteer tourism as a tourism activity incorporating volunteer activities" (p. 407). It is understood as the participants' activities of "enjoying a tourist experience with the benefit of helping others" (Tomazos & Cooper, 2012, p. 407). At the same time, Konstantinos Tomazos and Richard Butler (2009) emphasize that volunteer tourists "should not be confused with 'working tourists'" (p. 196), in that volunteer tourists are not paid, but rather they even pay to join the international volunteer activities of their choice. Citing earlier work by Claire Ellis (2003) and Stephen Wearing (2001), they argue as follows:

> A key element in the definition of a volunteer tourist is the absence of pay. Volunteer tourists do not get remunerated while on their trip. Instead they pay in a variety of ways for the privilege of volunteering. (Tomazos & Butler, 2009, pp. 196–197)

This argument has salience to the study presented here. The participants in my study are not necessarily "volunteer tourists" in its strict sense. Although they are engaged in a state-sponsored "volunteer program," the participants in this program are paid for their living allowances (*genchi seikatsuhi*), are given accommodations (*jukyo*), receive round-trip travel allowances (*ofuku tokohi*), and are provided with other benefits and supports including holidays (*kyuka*), allowances for domestic matters (*honpo shishutsu taio teate*), allowances upon finishing research (*kenkyu katsudo kanryo kin*), and access to various follow-up programs (JICA, 2022, pp. 9–10). Although the sponsor of the JLTs' program explains that these are "not salaries nor remuneration" but necessary support for living allowances (JICA, 2022, pp. 9–10), volunteer JLTs' lives are supported to some extent, and they do not "pay … for the privilege of volunteering" (Tomazos & Butler, 2009, pp. 196–197). In the meantime, a sense of "altruistic giving" (Tomazos & Cooper, 2012, p. 407) was mentioned both in the program overview by the JLTs' sponsoring institution, as well as by the volunteers in my interviews as an integral part of their activity (Motobayashi, 2015). Further, as illustrated below, expectations for a touristic experience are narrated by study participants as well. In this respect, the participants of the present study can be defined as volunteer tourists in the broadest sense, as they expected "enjoying a tourist experience with the benefit of helping others" (Tomazos & Cooper, 2012, p. 407) at least partially.

Altogether, the case in this study falls somewhere on a continuum between the volunteer tourism and working tourism mentioned above. Therefore, the abovementioned discussion on working tourists is quite informative in the analysis. In addition, the case could make a unique contribution to the discussion regarding the nature and definition of volunteer tourism through its very ambiguity and the questions it raises.

In short, the discussion on middling transnationals has evolved along with the conceptual development of "travelling workers," "working tourists," or "volunteer tourists" as way for scholars to understand the configuration of various activities such as work, leisure, and volunteering in mobility. Yet, "focused scrutiny of the way these tourists combine work and leisure" was still limited even in the 2010s (Jarvis & Peel, 2013, p. 114). In particular, more research is called for to focus on non-European and non-English speaking contexts, particularly in the context of volunteer tourism (Jarvis & Peel, 2013). In addition, language issues are underexamined in these studies, despite the theoretical capacity of language-focused analysis to broaden the scope of studies into work-leisure configuration by transnational movers (Motobayashi, 2020). To contribute to this line of inquiry, the study presented in this chapter focuses on language teachers' interests in the off-duty activities in volunteer destinations—in particular, the off-duty, foreign language learning opportunities.

Context of the Investigation: The *Nikkei* Volunteer Program

The "Nikkei Society Youth/Senior Volunteer program [*Nikkei shakai seinen/shinia borantia*]," which I will hereafter refer to as the "Nikkei volunteer program," is an international volunteer program organized by the Japan International Cooperation Agency (JICA). The Nikkei volunteer program is designed to support South American *Nikkei* communities, which are communities "of Japanese descent" or "related to Japan." JICA is an independent administrative agency dealing with the Japanese official developmental assistance (ODA) under the International Cooperation Bureau of Ministry of Foreign Affairs of Japan (MoFA). The Nikkei volunteer program was launched in 1996 and has since been operated under the umbrella of "JICA volunteers."

The JICA volunteers project consists of two international cooperation programs: the Nikkei volunteer program and the Japan Overseas Cooperation Volunteers (JOCV, or the Overseas volunteer program[1]). The relationship between these two programs and the position of the Nikkei program in the institution's administrative structure are rather complex. The "JICA volunteers," both in the Overseas and the Nikkei programs, are operated by the "Secretariat of 'Japan Overseas Cooperation Volunteers'" at the JICA Headquarters in Tokyo. In other words, the Nikkei program is operated under the name of the Overseas program. Indeed, in terms of scale, the Nikkei program is much smaller than the

[1] Hereafter in this chapter, as a way of reflecting the meaning implied in the Japanese names of these programs, and a means of focusing on the distinction between the "Nikkei" and the "overseas" spaces in Japanese, I will call the Japan Overseas Cooperation Volunteers the "Overseas" program, rather than using the abbreviation JOCV.

Overseas volunteer program.[2] In fact, the actual operations of the training sessions—an important component of the program—are outsourced to different organizations: the Japan Overseas Cooperation Association (hereafter JOCA) and the Association of Nikkei and Japanese Abroad [*Nikkeijin-kyokai*] (hereafter Nikkeijin Association) for the Overseas program and the Nikkei programs, respectively.

These differences between the two programs are due to their historical background. The Overseas program, officially abbreviated as JOCV in English, has a much longer history than the Nikkei volunteer program. It was created in 1965, about a decade later than Japan's participation in the Colombo Plan in 1954.[3] In terms of its aims and concept, the Overseas program is similar to the Peace Corps that was established in 1961 in the United States (JICA, n.d.). In other words, the Overseas program is more straightforwardly understood to be "international cooperation."[4]

On the other hand, the Nikkei volunteer program as its current form started in 1996 and its history is shorter. Behind the establishment of the Nikkei program in JICA in 1996 was a long history of Japanese emigration and social transformations both in Japan and in emigrants' destinations. Most immediately, the Nikkei volunteer program is situated in line with the history of Japan's post-WWII emigration and the Nikkei policies. As reported elsewhere (Motobayashi, 2015), Japan's emigration-related policies can be divided into three phases: "facilitation" of emigrant out-sending (mid-1950 to 1962), "support" for the Nikkei population in

[2] According to the data between 2006 and 2010, the Overseas volunteer program dispatched about 2000 volunteers per year on average (about 1400–1500 youth volunteers and 350–400 senior volunteers every year). To deal with this number of volunteers, recruitment was conducted twice a year and the training sessions were held four times a year. On the other hand, the Nikkei volunteers were about 40–50 per year (20–40 youth and 13–26 seniors, depending on the year). The recruitment and training sessions (a series of two) for them were once a year (see Motobayashi, 2015, for details).

[3] The Colombo Plan (The Colombo Plan for Cooperative Economic and Social Development in Asia and the Pacific) is an intergovernmental "cooperative venture for the economic and social advancement of the peoples of South and South-east Asia" that began operations in 1951 (The Colombo Plan Secretariat, n.d.), and in the context of Japan's international cooperation history, Japan's participation in 1954 is understood as Japan's first provision of technical cooperation to Asian countries (JICA, n.d.).

[4] Both the Nikkei volunteer program and the Overseas volunteer program have "youth" and "senior" categories, and thus, there are four categories in total: "Youth/Senior Overseas volunteers [*seinen kaigai kyoryoku tai/sinia kaigai borantia*]" and "Youth/Senior Nikkei volunteers."

their countries of residence (1962–1984), and "collaboration" with the Nikkei population (1985 to present). Along with the policy changes, Japanese emigration was reconceptualized in policy documents from "simple labor mobility" to "mobility of developmental capacity" in the early 1960s. In addition, Japanese emigrants and the Nikkei population began to be reconceptualized as bilingual and bicultural human resources based on the "need of diplomatic base" in state-to-state "interdependency" in the 1990s.

In this context, the linkage of the overseas emigration and international cooperation was facilitated, as embodied by the merger of Japan Emigration Services [*kaigai ijuu jigyo dan*] (JEMIS) (1963–1973) with the Overseas Technical Cooperation Agency [*kaigai gijutsu kyoryoku jigyodan*] (OTCA) (1962–1973) to create JICA in 1974. This occurred at the same time that emigration projects were shrinking as symbolized by the closure of the Overseas Emigration Council [*kaigai ijuu shingi kai*] (OEC) in 2001, which occurred as part of the administrative reform in the 1990s and 2000s and led to the discourse of a new era of Japan-Nikkei relations for "collaboration" (see Motobayashi, 2015, for a detailed historical analysis).

In this broad historical and institutional context, at the program level, a program called "Overseas Development Youths [*kaigai kaihatsu seinen*]" was launched in 1985 in the second decade after the JEMIS-OTCA merger; it continued for ten years until 1995 and changed into the "Nikkei volunteer program" in 1996. In short, behind the emergence of the Nikkei volunteer program in 1996, there was a historical continuation from the emigration-related policies of Japan and changes at the institutional and program operational levels. The program's change and relocation to and within JICA occurred in the mid-1990s as the result of a series of institutional restructuring and mergers, which reflected various changes at the level of macro policies due to social and political changes in Japan as well as in global society. Due to such a history, the Nikkei volunteer program is uniquely different from the JOCV in terms of destinations, position, scale, and the nature of expected volunteer services, including the meaning of Japanese language teaching.

The most important difference in light of the theme of this book is the meaning and value attached to Japanese language teaching. Japanese language teaching is framed quite differently within these two different programs. In the Nikkei volunteer program, the Japanese language teaching job is called "Nikkei Japanese Language School Teachers [*Nikkei nihongo gakko kyoshi*] (Nikkei JLSTs)," while the Japanese language teachers in the Overseas program are simply called "Japanese language teachers [*nihongo kyoshi*]." The targeted learner cohorts are also different: the Nikkei JLSTs in the Nikkei volunteer program are mostly expected to teach younger generations at various Nikkei Japanese Language Schools [*Nikkei nihongo gakko*] (Nikkei JLSs) operated by Nikkei Associations in South America; on the other hand, the Japanese language teachers in the Overseas program are mostly sent to secondary- and tertiary-level public institutions in ODA partner countries worldwide, mainly universities and colleges, and some middle schools.

In fact, the South American, local Nikkei JLSs have been understood as important sites for community maintenance, in particular during the initial stage of Japanese settlement in South America (see Motobayashi, 2015, Chap. 4 for more detail). Reflecting this history, perhaps, the ratio of the JLST position is higher than the ratio of Japanese language teachers in the Overseas program. As discussed elsewhere (Motobayashi, 2022), the expected contribution of the JLSTs volunteers is understood to be more than the teaching of Japanese. The Japanese language teaching in the Nikkei volunteer program is oriented toward a more holistic "emigration education" of the younger Nikkei population, being closely intertwined with the discourse of developing Nikkei identities. In the context of the volunteer training undertaken by the participants in this study, the discourse of heritage and heritage language was foregrounded. In this discourse, heritage education and heritage language were framed as desires held by Nikkei communities and cast the Nikkei population in line with Japanese national membership.

Methods

This study took an ethnographic approach, and data collection methods included participatory observation, document analysis, and in-depth interviews with the participants. The ethnographic observation and interview data were collected in Japan around 2010.[5] The participatory observation focused on the pre-departure training sessions of the Nikkei volunteer program members. As mentioned earlier, this study particularly focuses on the volunteers who applied for the Nikkei JLSTs jobs within the Nikkei volunteer program. I attended all the classes in the first training session with the volunteers and most of the lectures in the second training session, except for foreign language classes.

Semi-structured interviews were conducted with each participant, focusing on their overall trajectories such as their reasons for applying the Nikkei volunteer program, their expectations of the program, and their goals to be achieved or future plans. More specific questions on their choice of the program were also asked, including those regarding their choice of JICA's volunteer program, of the Nikkei program rather than the Overseas program, and of the Japanese teaching job category. Finally, their experiences of the volunteer trainings were asked about as well. Although I was not involved in any activities related to the JICA volunteer program before the research, my experience of studying applied linguistics in graduate schools in Japan and Canada affected my communication with the volunteers and staff members in the observation and interview processes. In addition, through my participation in the pre-departure training together with the participants, I felt that I developed a sense of rapport with my participants, which affected the ways in which we communicated and fostered a sense of ease.

Related documents were continuously collected before, during, and after the ethnographic data collection phase of the study in order to illuminate the historical and contemporary contexts in which this program has been situated at the intersection between international cooperation and the emigration policies of Japan, as well as to trace the administrative structure regarding the operation of this program in its current form. The concrete documents analyzed include the Diplomatic Blue Book of Japan; reports and proposals published by institutions such as MoFA, OEC, JICA, and the Nikkeijin Association; and pamphlets or promotional

[5] This study was conducted with a cohort of Nikkei volunteer candidates around 2010. Due to the small number of the volunteers, I do not identify the specific year in this study in order to ensure the confidentiality of the cohort members.

materials of the JICA and Nikkei volunteer programs. My discussion of the program in the previous section was based on a close examination of these documents.

Analysis: Off-Duty Expectations as Part of Volunteers' Decisions—Choosing a Destination, Choosing a Language

The volunteers' comments on their off-duty expectations appeared in the interviews in sporadic and unexpected ways, mostly in reaction to the researcher's question about the reason for the choice of destination. When applying to this volunteer program, candidates are required to indicate their three preferred program destinations. Being asked about the reason for the volunteer destination choice, the volunteers provided answers to explain how the particular locations fit their personal and professional interests—including their off-duty activities. Their off-duty interests at the volunteer destination were found in two forms in the study: foreign language learning and touristic experiences. This chapter focuses on the former.

Indeed, for some volunteers, the language spoken at the destination of choice was an important factor in the decision-making process. The language options offered by the Nikkei program were either Portuguese, for those headed to Brazil, or Spanish, for those being sent to countries such as Argentina, the Dominican Republic, or Paraguay. It should be noted that a unique feature of this volunteer program is that candidates are required to participate in a series of intensive language classes in the language of their destination as part of the pre-departure training sessions over a period of two months in Japan.

This section introduces the voices of three volunteers—Yuko, Keiko, and Tamaki—who, in their interviews, made explicit references to their interests in learning Spanish and Portuguese, respectively. Significantly, they linked the access to the foreign languages that this program affords—by immersion in an authentic linguistic community following pre-departure and systematic instruction—to their choice of destination and, ultimately, what they expected from the volunteer program *linguistically*—in addition to hoping to fulfil their Japanese language teaching responsibilities.

Keiko's Case: An "Unserious" Reason for Choosing Her Destination

The first volunteer introduced here is Keiko, who was in her 20s at the time of the research. Excerpt 1 is Keiko's response to the researcher's question on the reason for choosing the particular destination during the first interview.

Excerpt 1
Well, um, it's really an unserious reason, actually. Beaches are beautiful there. … *And, it's a Spanish-speaking country.* [Researcher: *Oh. Do you speak Spanish?*] *No, no, no. Because I thought Spanish would probably better broaden my future than Portuguese.* (Keiko, first interview)

Here, Keiko mentions touristic experiences and access to the Spanish language as something she expected outside of her work teaching Japanese. Mentioned first is touristic satisfaction with her volunteer destination, which, indeed, was a South American country famous for its resorts. Admitting half in jest that it may sound "unserious," Keiko referred to the beautiful beaches as the first reason for her choice of destination. This was followed by the second reason: the Spanish language spoken at the destination. Access to the local language was an important factor in Keiko's choice of destination ("And, it's a Spanish-speaking country"). Her reply ("No, no, no") in reaction to the researcher's question after her mention of the Spanish language ("Oh. Do you speak Spanish?") implies that she did not yet have Spanish language proficiency but foresaw an opportunity to learn the language through the volunteer program's predeparture training sessions and at the volunteer destination. In this comment, Keiko prioritizes Spanish over Portuguese based on her perception of the relative versatility of the language ("Because I thought Spanish would probably better broaden my future than Portuguese"). As such,

Keiko notes the possible extra activities, one touristic and the other linguistic, as reasons for her choice of volunteer destination.

Two other volunteers, Yuko and Tamaki, also commented on their choices of destination in relation to access to the local languages. Yuko indicated the same kind of prioritization of Spanish based on a scale-sensitive decision on language, while Tamaki prioritized the Portuguese language based on the specifics of her local employment in Japan.

Yuko's Case: "Spanish, if Possible" to Search for "Something Additional"

Yuko was another volunteer who was also in her 20s at the time of departure from Japan. She provided a similar perspective as Keiko's regarding the greater versatility, power, and merit of learning Spanish instead of Portuguese. However, she explained it in much more elaborated ways than Keiko, as part of her multi-tiered rationalization of prioritizing Spanish language learning over Portuguese ("Spanish, if possible"). In the Excerpt 2 and Excerpt 3 below, she explained that her choice of volunteer destination was based on the language spoken in the volunteer destination—that is, the Spanish language. Being asked about the choice of destination country, she replied as follows:

Excerpt 2

[T]he reason for [country name (her destination): a Spanish-speaking country in South America]. I made [the choice] all [name of the same country], from one to three [laughing] … *I was kind of interested in the Spanish language. I thought it would be good if I could speak Spanish* … That's a simple, you know, motivation. [Researcher: By the way, why Spanish, if I may ask?] Well yeah. Spanish is … when I was in England [to study English], *I had a friend who was Spanish*, that's one thing … *I thought it would be good if there was something additional to the Japanese … job in the country I am going to* … And then *I thought that if I become able to listen and speak Spanish, even a little, I would be able to have a conversation with that [Spanish] friend* … which is the *additional thing* I would get … *Actually, Portuguese is good, too, as I had a Brazilian friend as well* [laughing] ... But, *I guess I would prefer speaking Spanish*, because it is also *spoken in wider areas*, you know ... People say both Portuguese and Spanish are *the languages needed in Japan now*. So ... I thought either would be okay because *both are useful ... But it was like, Spanish, if possible* [laughing]. (Yuko, first interview, emphasis added)

In Excerpt 2, Yuko demonstrates her rationalization for the choice of a Spanish-speaking country. She uses a two-tier rationalization for prioritizing the Spanish language and a Spanish-speaking country. At the beginning of Excerpt 2, she indicates an interest in Spanish ("I was kind of interested in the Spanish language" [Excerpt 2; Line 3–4]) and a willingness to learn and speak it ("I thought it would be good if I could speak Spanish" [Excerpt 2; Line 4]). Then, the first rationale for the choice of Spanish-speaking country is stated: her previous socialization with a Spanish speaker in a study abroad program in England. She "had a friend who was Spanish" (Excerpt 2; Line 7) and she "thought that if [she] become[s] able to listen to and speak Spanish, even a little, [she] would be able to have a conversation with that [Spanish] friend" (Excerpt 2; Line 9–11). This indicates her understanding of the Spanish language as a means of intercultural socialization with linguistic and cultural others associated with the Spanish-speaking world. However, as she admits herself, having a friend who speaks the language could also be a motivation to learn Portuguese, because she had a Brazilian friend in England as well (Excerpt 2; Line 12). Thus, her previous contact and personal friendship with the language's speaker could not fully legitimize her choice of the Spanish-speaking country.

Following this, Yuko moves on to the second tier of her rationalization for prioritizing Spanish: the greater versatility it offers as a means of intercultural communication. While she acknowledges the value of both Portuguese and Spanish in terms of intercultural socialization at personal levels, as well as each language's importance in an increasingly multilingual and diverse Japan (Excerpt 2; Line 14–16), she places Spanish and Portuguese on a scale of merit in terms of the range of their uses and judges them based on her understanding of the relative versatility and value of each language ("I guess I would prefer speaking Spanish, because it is also spoken in wider areas, you know" [Excerpt 2, Line 13–14]).

Finally, Yuko decides to go with "Spanish, if possible" (Excerpt 2; Line 16–17), prioritizing Spanish in light of its perceived level of global versatility.

Interestingly, versatility and commonality are the different sides of the same coin, as Yuko illustrates in her interview using the case of the English language. Excerpt 3 is taken from my second interview with Yuko, in which we were triangulating the content of the first interview and Yuko further elaborated on the value of the Spanish language. She said:

Excerpt 3

[I]t's so common that someone can speak English, you know. So in that sense as well, I thought [it would be good] if I could learn something additional at the same time ... in participating in this ... Not just the activity as a volunteer, but additionally, as language learning, like something that I could use in the future after I come back to Japan. It would be great if I could acquire something additional in some senses, you know, it's natural to think that way, isn't it [laughing]? ... I mean, I thought things like that, and also, *Spanish is actually spoken in a wider range of areas ... The opportunities to use it are [wider], yeah.* (Yuko, second interview, emphasis added)

In Excerpt 3, Yuko mentions the even greater versatility and utility of English over Spanish, but rather in a negative way, as she describes English proficiency as being too "common" ("it's so common that someone can speak English" [Excerpt 3; Line 1]). Her noting of the "commonality" of having English proficiency better highlights the value of the Spanish language as an additional foreign language; she suggests it may have more distinct value than English. Indeed, Yuko is one of those volunteer participants who had invested in English study abroad prior to joining the volunteer program as a JLT (Motobayashi, 2015). Based on the perception of the "commonality" of English, she is now exploring the possibility of acquiring Spanish as a second or third foreign language and taking the volunteer period as an opportunity to do so.

These interview excerpts show a multi-tiered rationalization toward the decision on language (and destination along with it) in light of participants' sense of the relative value of different languages. A careful comparison and calculation, which is rank- and scale-sensitive, was found before participants reached a decision to prioritize Spanish over Portuguese in their applications to this program. This is best expressed in the phrase, "Spanish, if possible" (Excerpt 2; Line 16–17), which was based on Yuko's perspective regarding the relative versatility of the Spanish language in the global context, which she noted in both the first interview ("I guess I would prefer speaking Spanish, because it is also spoken in wider areas" [Excerpt 2; Line 13–14]) as well as the second ("Spanish is actually spoken in a wider range of areas … The opportunities to use it are [wider], yeah" [Excerpt 3; Line 8–9]). The value of the Spanish language as having a distinctive value as an additional foreign language besides English is emphasized as well; and the language learning is conceptualized as "something additional to the Japanese … job in the country [she was] going to" (Excerpt 2; Line 8–9).

Foreseeing, then, the language learning opportunities provided by the Nikkei volunteer program that sends them to South American destinations, both Keiko and Yuko compared the two European languages accessible through the volunteer program in terms of their relative versatility and future utility. Based on an understanding that Spanish "is more widely spoken" (Yuko) in the world and therefore "would better broaden my future than Portuguese" (Keiko), they considered learning Spanish to be a better investment in a more versatile resource than Portuguese.

One thing to be noted is the way both Keiko and Yuko envisioned the Spanish language and its space. Yuko's mobility to South America was rationalized with her willingness to access the Spanish language based on her having a "Spanish" friend from her former cosmopolitan experience of an English study abroad program. Here, her mobility to the volunteer destination country is imagined, at least partially, as participation in "a linguistic sphere" (i.e., the "Spanish-speaking sphere") that transcends the particular geographic location in South America and stretches on a global scale to include both the South American volunteer destination and the friend who is Spanish. Keiko's mention of the global versatility of the Spanish language in Excerpt 1, although being rather brief, also suggests that she perceives the Spanish language as a language transcending the specific volunteer destination of her choice in South America. In these cases, Spanish is understood as not exclusively the language of the destination country but as a language with some sort of cosmopolitan spatiality. Keiko and Yuko's choices seem to be based on an imagination of a space inhabited by Spanish speakers as well as the value of the language in relation to their own futures, when they, too, will participate in a cosmopolitan linguistic sphere.

Tamaki's Case: The Value of Portuguese in Her Own Employment Context in Japan

Yuko and Keiko's reference to the "value" of Spanish is rather vaguely oriented to intercultural socialization and cosmopolitan spatialization, accompanied by a careful consideration of languages' relative scalar ranks. On the other hand, Tamaki, another study participant, commented that the Portuguese language—the language accessible in Brazil in this program—was a better resource for her to volunteer in a different context than Keiko and Yuko.

Tamaki had worked as a part-time Japanese second language teacher in a region of Japan with a large Nikkei-Brazilian community before her application to this program. She had been involved in the education of Nikkei-Brazilian students in Japan since she was an undergraduate student, when she started volunteer teaching for those students. She continued her studies in a graduate program and had been "just passionate about Brazil since then" (Tamaki, first interview). In fact, her participation in the Nikkei program at the time of this research was after her second attempt to enter the program, as she had not been accepted on her first attempt immediately after finishing graduate school. Given that history before she became a member of the Nikkei volunteer program, Tamaki had forged her perspective on the acquisition of the Portuguese language and gaining experiences in Brazil, which she explained as follows:

Excerpt 4

Around the end of [my] undergrad, I started the volunteer teaching for children. And I found it again that I like kids, and that it's fun, and I thought I would get the teacher certificate; like that's the time I was thinking like that. Then, *I thought, I would like to acquire Portuguese that becomes useful once I get back to the school teaching*, and also [I thought] *the experiences in Brazil would probably be acknowledged.* It was things like that. (Tamaki, first interview, emphasis added)

In the above excerpt, Tamaki refers to the value of Portuguese in more concrete terms in light of the particular context and experiences of her work in Japan. Tamaki is aware that Portuguese would "[become] useful once [she got] back to the school teaching" ("I thought, I would like to acquire Portuguese that becomes useful once I get back to the school teaching" [Excerpt 4: Line 4–5]) and her "experiences in Brazil would probably be acknowledged" (Excerpt 4: Line 6) in contexts such as the school where she had worked. In other words, she conceptualizes the value of her prospective language learning at the volunteer destination as linked to an employment-oriented value when she returns to work in Japan.

Her perspectives on the value of Portuguese and of authentic experiences in Brazil are further elaborated upon in Excerpt 5. This is where she explains why the Nikkei program was a better choice for her than the Overseas program, the other volunteer program operated within the JICA volunteer scheme but unrelated to the South American Nikkei issues.

Excerpt 5

I thought it would become a good part of my career [*kyaria ni naru*] when I came back, if I went to Brazil [with the Nikkei program rather than going to the Overseas programme]. … *It's* [a matter of] *career, you know. I would be able to understand the locally spoken Portuguese, at least to a little extent.* And well, um, *the children I was taking care of as a volunteer* [at the time when she was thinking of applying to the Nikkei program for the first time] *were Brazilian children, as a matter of fact, so* [I thought], *like, what is the country like where those children were born? And, I also used to teach Japanese to the parents, to adults.* So when I saw people like them, well, um, I thought, like, I wanted to feel the difference between [Brazil and] Japan, kind of. Then I should, like, go to their country. And so, from early on, I had decided that the destination should be Brazil … I wanted to go to [Brazil]. (Tamaki, first interview, emphasis added)

Here, Tamaki explains her perspective on the choice of Brazil as her desired destination. Using the expression "it would become a good part of my career [*kyaria ni naru*]," she elaborates on how the authentic experience in Brazil could be valuable in light of her career goals ("I thought it would become a good part of my career [kyaria ni naru] when I came back, if I went to Brazil" [Excerpt 5: Line 1–2]). She also saw Portuguese language skills as an asset for future career opportunities, as she had previously worked with Nikkei-Brazilian children in Japanese schools ("It's [a matter of] career, you know. I would be able to understand the locally spoken Portuguese, at least to a little extent" [Excerpt 5: Line 3–5]).

In this excerpt, Tamaki demonstrates that she understands her prospective mobility to Brazil through the Nikkei program as a way to access the authentic Portuguese language as well as to gain authentic cultural experiences and knowledge of Brazil. These goals and expectations were closely connected to her prospective career trajectory and professional identity as an educator for Nikkei-Brazilian students in Japan. Tamaki's choice of Brazil as her volunteer destination and her interest in the Portuguese language and the authentic experiences accessible in Brazil were situated within her career prospects. These prospects link her previous job before the volunteer work to the prospective volunteer activities in Brazil as well as her future career plans when she returns to Japan.

In Tamaki's case, the Portuguese language is imagined in a much more concrete way, being linked to the locality and particularity of Brazil, Brazilian people, and a context based on her prior contact with Nikkei-Brazilian children and their parents through real-life teaching experience. It can be contrasted to a rather abstract description by Keiko and Yuko of Spanish as a cosmopolitan, versatile language.

Discussion: Shedding Light on the Linguistic Aspect of Middling Transnationals' Subjectivities

The above findings can be read as concrete cases of middling transnationals' subjectivities and their work-leisure configuration practices. The narratives of the three volunteer JLTs show part of their efforts for a configuration of various activities, as each tried to make her overseas stay economically sustainable and rewarding, as well as experientially fulfilling and meaningful. This chapter thus adds to the discussion of middling transnationals (Conradson & Latham, 2005a, 2005b; Thompson & Tambyah, 1999), working tourists, or travelling workers (Jarvis & Peel, 2013; Uriely, 2005), by shedding light on the linguistic aspects of middling transnationals' subjectivity and agency. The data presented in this chapter show volunteer JLTs' efforts to proactively and strategically manage language-related resources (De Costa et al., 2016). In so doing, they produce and reproduce ideologies of named and objectualized languages. These ideologies comprise part of their decisions and expectations for upcoming work-leisure configurations—configurations that are intertwined with their imaginations of their transnational mobility to come.

A combination of linguistic inquiry and middling transnationalism studies—as I have attempted in this chapter—allows us to capture the nuanced but important ways in which language ideologies played a role in these study participants' decision-making regarding mobility. In narrating their plans, expectations, and aspirations respective to their prospective mobilities, the participants demonstrated their own rationalization and careful consideration of their choices prior to applying to the Nikkei program and departing overseas. Although the "off-duty" activities—both touristic expectations and language-related expectations—were mentioned as "additional" (Yuko) to the primary job and associated with somewhat "unserious" (Keiko) reasons, the expected off-duty activities constituted an important part of all three study participants' decision-making and the rationalization behind their choices, including the choice of destination among the options of the South American locations available through the Nikkei volunteer program.

The extent to which the three volunteers succeeded on their intentions is unclear as of this writing, and whether they actually attempted to learn new languages at their destination sites is unknown; a follow-up study is planned. What has been shown in this study, however, is that possible access to foreign language learning (be it either for leisure-oriented or career-oriented purposes) and other experiences (such as tourist experiences) did have an impact on the volunteers' decision-making prior to departure, in addition to JLT career-oriented imperatives. The findings presented in this chapter regarding work-leisure configurations are in line with those reported as practices that distinguish middling transnationals. Future research will address how the discussion in applied linguistics around resource-oriented views on language (Motobayashi, 2021) makes a unique contribution to the argument of middling transnationals; it could help explain the transnationals' amalgamation of the economic and the experiential at cultural and linguistic levels.

References

Bauman, Z. (1998). *Globalization: The human consequences*. Polity Press.

Clarke, N. (2004a). Free independent travelers? British working holiday makers in Australia. *Transactions of the Institute of British Geographers, 29*(4), 499–509. https://doi.org/10.1111/j.00202754.2004.00144.x

Clarke, N. (2004b). Mobility, fixity, agency: Australia's holiday working plan. *Population, Space and Place, 10*(5), 411–420. https://doi.org/10.1002/psp.347

Clarke, N. (2005). Detailing transnational lives of the middle: British working holiday makers in Australia. *Journal of Ethnic and Migration Studies, 31*(2), 307–322. https://doi.org/10.1080/1369183042000339945

Clifford, J. (1992). Traveling cultures. In L. Grossberg (Ed.), *Cultural studies* (pp. 96–116). Routledge.

Collins, F. L. (2014). Teaching English in South Korea: Mobility norms and higher education outcomes in youth migration. *Children's Geographies, 12*(1), 40–55. https://doi.org/10.1080/14733285.2013.851064

Conradson, D., & Latham, A. (2005a). Transnational urbanism: Attending to everyday practices and mobilities. *Journal of Ethnic and Migration Studies, 31*(2), 227–233. https://doi.org/10.1080/1369183042000339891

Conradson, D., & Latham, A. (2005b). Friendship, networks and transnationality in a world city: Antipodean transmigrants in London. *Journal of Ethnic and Migration Studies, 31*(2), 287–305. https://doi.org/10.1080/1369183042000339936

De Costa, P., Park, J. S.-Y., & Wee, L. (2016). Language learning as linguistic entrepreneurship: Implications for language education. *Asia-Pacific Education Researcher, 25*(5–6), 695–702. https://doi.org/10.1007/s40299-016-0302-5

Duncan, T. (2008). The internationalisation of tourism labour markets: Working and playing in a ski resort. In T. Coles & C. Hall (Eds.), *International business and tourism: Global issues, contemporary interactions* (pp. 181–194). Routledge.

Ellis, C. (2003). Participatory environmental research in tourism: A global view. *Tourism Recreation Research, 28*(3), 45–55. https://doi.org/10.1080/02508281.2003.11081416

Haverig, A., & Roberts, S. (2011). The New Zealand OE as governance through freedom: Rethinking "the apex of freedom". *Journal of Youth Studies, 14*(5), 587–603. https://doi.org/10.1080/13676261.2011.565042

Ho, E. (2011). Migration trajectories of "highly skilled" middling transnationals: Singaporean transmigrants in London. *Population, Space and Place, 17*(1), 116–129. https://doi.org/10.1002/psp.569

Jarvis, J., & Peel, V. (2013). Tourists for hire: International working holidaymakers in a work based destination in regional Australia. *Tourism Management, 37*, 114–124. https://doi.org/10.1016/j.tourman.2012.10.014

JICA. (2022). JICA Overseas Cooperation Volunteers: Program overview [*JICA kaigai kyoryoku tai: jigyo gaiyo*]. Retrieved May 10, 2022, from https://www.jica.go.jp/volunteer/outline/publication/pamphlet/pdf/gaiyo.pdf

JICA. (n.d.). History of JICA Volunteer programs [*JICA borantia jigyo no ayumi*]. Retrieved December 20, 2022, from https://www.jica.go.jp/volunteer/outline/history/index.html

Juffermans, K., & Tavares, B. (2014). Trajectories and repertoires as conceptual tools for a sociolinguistics of globalization. *STAR Project, Research brief #3.* Retrieved November 11, 2015, from http://wwwen.uni.lu/recherche/flshase/education_culture_cognition_and_society_eccs/research_institutes/research_on_multilingualism_mling/projects_and_publications/star_sociolinguistic_trajectories_and_repertoires_luso_luxo_african_identifications_interactions_and_imaginations#

Kennedy, P. (2010). Mobility, flexible lifestyles and cosmopolitanism: EU postgraduates in Manchester. *Journal of Ethnic and Migration Studies, 36*(3), 465–482. https://doi.org/10.1080/13691830903426838

Lash, S., & Urry, J. (1994). *Economies of signs and space.* Sage. https://doi.org/10.4135/9781446280539

Motobashi, K. (2015). *Language teaching as foreign policy: Japanese language teachers in Japan's international cooperation volunteer program.* [Doctoral dissertation, University of Toronto]. Tspace Repository.

Motobayashi, K. (2016). Language teacher subjectivities in Japan's diaspora strategies: Teaching my language as someone's heritage language. *Multilingua, 35*(4), 441–468.

Motobayashi, K. (2020). *Mobility, identity, and the social turn in applied linguistics: Theoretical and pedagogical implications*. Invited symposiast at the ESRC-JSLARF Symposium, October 3, 2020.

Motobayashi, K. (2021). *Unsettling capital in 'investment' and 'consumption'*. Paper presented on the panel "unsettling capital and language", organized by W. Simpson, & J. O'Regan. Sociolinguistic symposium 23, June 9, 2021.

Motobayashi, K. (2022). Discursive construction of heritage desire: Nikkei identity discourse in a layered politics of representation. In M. Mielick, R. Kubota, & L. Lawrence (Eds.), *Discourses of identity: Language learning, teaching, and reclamation perspectives in Japan*. Palgrave Macmillan.

Munt, I. (1994). The other postmodern tourism: Travel, culture and the new middle class. *Theory, Culture and Society, 11*(3), 101–124. https://doi.org/10.1177/026327694011003005

Raj, D. (2000). *Where are you from? Middle-class migrants in the modern world*. University of California Press.

The Colombo Plan Secretariat. (n.d.). *The history of the Colombo Plan*. Retrieved March 6, 2023, from https://colombo-plan.org/cp_about/

Thompson, C. J., & Tambyah, S. K. (1999). Trying to be cosmopolitan. *Journal of Consumer Research, 26*(3), 214–241. https://doi.org/10.1086/209560

Tomazos, K., & Butler, R. (2009). Volunteer tourism: The new ecotourism? *Anatolia, 20*(1), 196–212.

Tomazos, K., & Cooper, W. (2012). Volunteer tourism: At the crossroads of commercialisation and service? *Current Issues in Tourism, 15*(5), 405–423. https://doi.org/10.1002/1522-1970(200101/02)3:13.3.CO;2-D

Uriely, N. (2001). "Travelling workers" and "working tourists": Variations across the interaction between work and tourism. *International Journal of Tourism Research, 3*(1), 1–8. https://doi.org/10.1002/1522-1970(200101/02)3:1%3C1::AID-JTR241%3E3.0.CO;2-M

Uriely, N. (2005). The tourist experience: Conceptual developments. *Annals of Tourism Research, 32*(1), 199–216. https://doi.org/10.1016/j.annals.2004.07.008

Wearing, S. (2001). *Volunteer tourism: Experiences that make a difference*. CABI. https://doi.org/10.1079/9780851995335.0000

Wilson, J., Fisher, D., & Moore, K. (2010). The "OE" goes home: Cultural aspects of a working holiday experience. *Tourist Studies, 9*(1), 3–21. https://doi.org/10.1177/1468797609360590

Part II

Language-Motivated Voluntourism as Precarious Labor

6

Dreaming of Entrepreneurship, Europe, English, and Freedom: Voluntourism as a Pure Survival Strategy

Larissa Semiramis Schedel

Language Tourist or Illegalized Migrant?

Issam[1] is a young man in his 20s from the Maghreb (Northwest Africa). He travels across Europe without a residence permit and lives from volunteer jobs. On his journey, he (in)formally learns different languages to differing degrees. When I thank him for sharing his story with me after several interviews about his language learning experiences, he replies that he usually does not tell anybody about his life in "illegality" because people could look down on him. When people ask him what he does, he answers that he travels to learn Spanish and he waits for his residence permit. He adds that people from European countries do not know the

[1] I use this name and omit his nationality to anonymize the identity of my research participant.

L. S. Schedel (✉)
University of Bonn, Bonn, Germany
e-mail: lschedel@uni-bonn.de

L. S. Schedel, C. Jakubiak (eds.), *Voluntourism and Language Learning/Teaching*, Palgrave Advances in Language and Linguistics,
https://doi.org/10.1007/978-3-031-40813-7_6

real meaning of "residence" and that they are unaware of the freedom their passports grant them.

During my fieldwork in the language tourism industry, I met many international young people who engaged in volunteer work to learn a language (see also Schedel, 2022)—and several of them were in the middle of the application process for a residence permit.[2] Issam's illegalized status distinguished him from most of them as his visa had expired. His story is yet far from being a single case. His voluntourism practices and language learning trajectory resonate with what I have heard in other interviews and what I could observe in the field. However, his story contradicts the usual representation of voluntourists in the literature and therefore merits a closer look. It invites us to question common assumptions about voluntourism and allows us to explore the social consequences of the education-tourism-migration nexus.

In this chapter, I use the terms "illegal/ized" not with the aim to criminalize voluntourists without residence documents but rather to point out their vulnerable status. I inscribe my study in a canon of research that analyzes "illegality as a 'phenomenon' rather than as a problem, exploring its nature, origins, processes, perceptions, experiences, and impacts rather than searching for solutions" (Ruhs & Anderson, 2010, p. 196). I understand "illegality" as a variable construct produced by nation-states to delimit mobility and regulate access to resources and rights, thereby distinguishing human beings as (un)desired, (un)deserving, (il)legal migrants/refugees/workers. These binary labels become political techniques for managing people on the move and affect how they are received in the host society (Hamlin, 2021).

The chapter starts with a literature review on language learning in the context of voluntourism as a steppingstone to labor migration. After explaining my ethnographic and speaker-centered approach, I explore the (language) desires and ideologies that undergird Issam's imagination of Europe, his language learning trajectory, and his voluntourism

[2] Among the voluntourists working with Issam in Italy were, for instance, also young people from South American countries who had Italian ancestors that made them eligible to acquire Italian citizenship. Voluntourism allowed them to lower the costs of their stay in Italy during the application process—as they were not allowed to work yet—and to learn Italian.

practices. In the conclusion, I discuss the neoliberal governance of his transnational mobility and call for more research on voluntourism "off-the-beaten-track."

Voluntourism and Language Learning as Entrance to Labor Migration

"Volunteer tourism" or "voluntourism" was first described 20 years ago by Stephen Wearing (2001) in his often-cited work, *Volunteer Tourism: Experiences that Make a Difference*. He defined voluntourists as

> those tourists who, for various reasons, volunteer in an organized way to undertake holidays that might involve aiding or alleviating the material poverty of some groups in society, the restoration of certain environments, or research into aspects of society or environment. (Wearing, 2001, p. 1)

The public face of voluntourism and its representation in most of the contemporary scholarly literature aligns with this definition, depicting it as a leisure activity of young White middle-class people from the Global North who engage as volunteers in humanitarian aid projects in the Global South for altruistic reasons and return home after some time (some weeks, months, or up to a year). The roots of international voluntourism are indeed intertwined with the rise of "development" work (Baillie Smith & Laurie, 2011; Butcher & Smith, 2015), and still today, many voluntourism practices are, as Isabel Remers (2022) argues, an extension of colonialism and missionary practices disguised as Western aid.

However, as discussed in the introduction of this volume, voluntourism has evolved into a highly segmented industry (Stainton, 2016) and is not limited to North-South mobilities or development and conservation work, but includes also North-North/South-North/South-South mobilities (e.g., Baillie Smith et al., 2018; Brennan, 2018; Dlaske, 2016), non-altruistic/-moral motivations (Coghlan & Fenell, 2009; McGloin & Georgeou, 2016), hosts that are for-profit organizations, occupations in varying sectors, as well as costs or compensations ranging from tourists having to pay for the work experience (Mostafanezhad, 2014) to tourists getting paid for their volunteer work (Baillie Smith et al., 2022). Critical scholars have also deromanticized voluntourism by unveiling its commodification and neoliberalization (e.g., Lyons et al., 2012; Mostafanezhad, 2014; Vrasti & Montsion, 2014).

The neoliberalization of voluntourism experiences entails, among other things, their framing and promotion as an educational mobility with numerous opportunities for enhancing one's skills (Lyons et al., 2012). Voluntourists are often well-educated (Tomazos & Butler, 2009) and engage in voluntourism to further boost their CVs, hoping that the work experience and the language or other skills acquired abroad might function as a marker of distinction on a competitive globalized labor market (e.g., Choi, this volume; Jang, this volume; McGloin & Georgeou, 2016). Several studies have given examples where language learning has been a (key) motivational factor for tourists to engage in volunteer work abroad (e.g., Brennan, 2014, p. 262; Jang, 2017, pp. 136–139; Schedel, 2022; Takahashi, 2013, p. 68; Tiessen, 2018, p. 35), and voluntourism has, in fact, evolved into an important product of the booming immersion experience industry.

A look at current offers on voluntourism agencies' websites shows that many voluntourism providers respond to the voluntourists' quest for language learning opportunities and promise language immersion in exchange for some hours of "help" (or rather labor). Living and working in an environment where the target language is spoken is assumed to be the most effective learning method and presumably allows for becoming fluent in this language very quickly. However, research has found that not every kind of volunteer job is equally suitable for learning a language (e.g., Jang, 2017; Schedel, 2022).

A common motivation to engage in language learning through voluntourism abroad is also rooted in the wish to achieve, in the long term, permanent residence and to work in the host country (Ghazarian 2021, p. 326; Vrasti & Montsion, 2014), often necessitating language skills on a specific level. Or, at times, staying in the host country might not necessarily have been planned in the first place, but voluntourism jobs might function as entry to a regular occupation (see also McDaid & Sunyol, this volume).

While most of the voluntourists go the legal way and either do not need a visa (e.g., Europeans in Europe) or hold a valid visa, there are numerous media reports about third country nationals who (mis)used (language) student visas as a gateway to Europe, the USA, Australia, or New Zealand, meaning they entered the nation-state with this visa, but overstayed it or worked without a permit during their stay. (Unserious) language schools and travel agencies are the accomplices and beneficials of this visa "fraud" as they sell (or at times even only pretend to sell) the language travel packages for which the visas get issued and do not usually repay the clients if the visa or its extension is denied. Businesses also profit from voluntourists (legalized and illegalized ones—hosts might not even know about their status, see Ruhs & Anderson, 2010), employing them as a temporary, cheap workforce to fill labor shortages in unskilled, low-wage jobs (e.g., Brennan, 2018; Reilly, 2015; Schedel, 2022), thereby de-skilling these well-educated but often work-inexperienced young people (McDaid & Sunyol, this volume). Furthermore, their learner status of the local language is often used to hierarchize international voluntourists and assign them tasks that "native"-speaking voluntourists or co-workers are not eager to do (Schedel, 2022).

While student or tourist visas might but do not necessarily include work permits, voluntourism constitutes a grey area for those who wish to work without a permit to finance or at least to lower the costs of their stay abroad, as volunteering does not legally count as work but grants access to an accommodation, a meal, and/or a small renumeration. Yet, at the same time, voluntourists, and especially illegalized ones, are not protected by labor rights and are therefore particularly vulnerable workers (Brennan, 2018; Gomberg-Munoz, 2016; Reilly, 2015).

So far, the phenomenon of such a language learning-motivated and migration-oriented voluntourism has not received much scholarly attention. Especially, the voluntourism practices of illegalized migrants have not been investigated in detail yet. Furthermore, literature on language-related educational mobilities has focused heavily on American and European language learners (for a critique, see Kinginger, 2009) and only recently on learners from Asia. However, simply because language learners and/or voluntourists from African countries rarely appear in the literature does not mean that they do not exist. By giving voice to Issam, this contribution aims at revealing the subjectivity and experiences of such a marginalized language learning- and migration-motivated voluntourist to reassess common beliefs about voluntourism and language learning abroad.

Methods

This chapter draws on several biographical in-depth interviews[3] conducted with Issam over the course of five years and on ethnographic data (observations, interviews, promotional material, job advertisements, etc.) from different hostels employing voluntourists. The study is part of a larger multi-sited ethnographic research project named “Language skills for sale” (2018–present) that investigates the commodification of English and its social consequences in the context of the booming language tourism industry in the bilingual (with Maltese and English) island state of Malta from a critical sociolinguistic perspective (Heller, 2002).

When I first met Issam in Malta in April 2018, we were both staying in the same hostel, where many international English learners either worked as volunteers and/or stayed while attending English classes at one of the local language schools. The latter was true for Issam. As I was interested in the motives for and experiences of language tourism, I asked him to do an interview and we have stayed in contact ever since. In summer 2019, I visited him in Italy, where he was volunteering in a hostel, and he kept me informed about his subsequent trajectory and experiences via WhatsApp audio and text messages. We did another online interview in winter 2021 and one in summer 2022, when he was volunteering in Spain.

I use a speaker-centered approach that I combine with the ethnographic approach to contextualize the biographical narrative from his interview data (Flubacher, 2022). Speaker-centered approaches are

[3] We conducted the interviews mostly in French, a language in which we both felt comfortable. Occasionally, we switched to English. This chapter written in English conceals the highly multilingual experience that Issam lived during his journey across Europe as well as the circumstances under which this research has been conducted (see Meyer Pitton & Schedel, 2022, for an attempt to unveil the politics of language choice in this research setting).

typically used to explore language biographies or specific language learning trajectories, "to get an understanding of experiences that are usually deemed different" (Flubacher & Purkarthofer, 2022, p. 9) and to give voice to marginalized/minoritized subjects. More precisely, they give insights into how speakers "conceive of, live with, experience, dream of languages and whatever it is they consider as language(s)" (Flubacher & Purkarthofer, 2022, p. 6), as well as how they navigate dominant discourses and position themselves in structures of power and inequality. An ethnographic speaker-centered approach further allows for the exploration of affect and emotions (Busch, 2017; Flubacher, 2022) that have only recently received attention in research on language learning (see, e.g., Doerr & Taïeb, 2020; Kinginger, 2004; Kramsch, 2006; Piller & Takahashi, 2006).

The analytical concept of "language desire" is an attempt to bring this "neglected dimension" of language learning (Kramsch, 2006; Motha & Lin, 2014) into the focus of the research. A desire for a specific language is a complex multifaceted construct that usually comes along with other desires and is dialectically constituted through macro-discourses (which are, for instance, spread in the media) and individual experiences (Piller & Takahashi 2006, p. 59). I draw on the concept of "language desire" to explore Issam's aspiration to learn English and other languages during his journey through Europe, that is, the imaginaries that he attaches to these languages, the identity he hopes to achieve through learning these languages, as well as the things he thinks having a good command of these languages will allow him to do (see also Sinervo, this volume, on "linguistic dreaming").

Issam's Dreams

As I am using the concept of (language) desire as an entry point to make sense of Issam's narrated journey, I have structured the analysis around his dreams—how they evolve, shape, and are shaped by his voluntourism practices.

Dreaming of Entrepreneurship, Europe, and English: Language Tourism in Malta

Issam was born and raised in a big town in the Maghreb. At home, he spoke a local variety of Arabic and he was proud of his "accent-free" French—which he learned and used at school—and that people thought he was from France when they heard him talking. He had completed a bachelor's degree in accounting, obtained a master's degree in supply management, and gained some work experience in the finance and the logistic sector. However, these jobs did not satisfy him. Since he was little, Issam had always dreamed of becoming an entrepreneur, of being a director of a company, of being his own boss. He watched lots of YouTube videos to train himself—videos about how to start your own business and about the life of popular billionaires; he wanted to study their entrepreneurial mindsets. Inspired by these "success stories" (Park, 2010), he aspired to start his own e-commerce business, but not in his home country. He instead dreamed of undergoing further training and starting his business in Europe, where the education opportunities and economic conditions for entrepreneurs seemed much better.

During the interviews with me in French, he used many business terms in English. The use of a technical vocabulary in English was, however, not the only thing that struck me about his narrative. Issam had totally internalized the neoliberal discourse of the entrepreneurial self who had to self-optimize himself (Bröckling, 2007), and the improvement of his English skills—the language of business as he said—was for him the first step to achieve his dream of becoming an entrepreneur in Europe. In my first interview with him, he said the following about the relevance of English:

> Issam: C'était vraiment nécessaire pour moi d'apprendre l'anglais parce que c'est une leng- c'est une langue qui est très importante, c'est une langue qui est universelle. [respire] Et sans l'anglais, tu pourrais pas trouver un très bon emploi, tu pourras pas voyager, tu pourras pas faire vraiment autant de choses. Du coup, voilà, l'anglais c'est, c'est une langue qui est très importante. (…) Sans l'anglais, tu pourrais pas faire énormément de connaissances, euh tu pourras pas travailler. Par exemple moi aussi, vu que je suis

intéressé par tout ce qui est entreprena-, tout ce qui est dropshipping et business. Donc la majorité des entrepreneurs, les tops, ce sont des Américains. Du coup, leurs formations, leurs séminaires, c'est en anglais. Du coup, tout est en anglais maintenant. Quasiment tout.

My translation:

Issam: It was really necessary for me to learn English because it's a lang- it's a very important language, it's a universal language. [inhales] And without English, you wouldn't be able to find a very good job, you won't be able to travel, you really won't be able to do lots of things. Thus, well, English that's that's a language that is very important. (...) Without English, you wouldn't be able to meet many people, uhm you won't be able to work. Me too, for instance, as I am interested in everything that is related to entreprene-, everything that is related to dropshipping and business. Well, most of the entrepreneurs, the top ones, are Americans. Thus, their training, their seminars, it's all in English. So, everything is in English now. Nearly everything.

This interview excerpt reveals the linguistic ideology of English as a global language and the key to social inclusion and material success in a globalized world (Park, 2011). Issam's desire for English was closely intertwined with his desire of becoming an entrepreneur and with a desire for (participating in) the Western world for which English functions as index. He was convinced that if he wanted to learn from "the best," he needed to invest in his English skills. This linguistic investment can be understood as an act of "linguistic entrepreneurship", that is, "the act of aligning with the moral imperative to strategically exploit language-related resources for enhancing one's worth in the world" (De Costa et al. 2016, p. 696).

At college, Issam had learned some English, but only 1 h/week. In 2010, he took some language classes proposed by the British Council (3 h/week) and in 2017 by an American language school in his hometown. However, despite these efforts, his English skills were still less than he desired. Issam deplored the grammar-focused teaching style of these classes and was convinced that the short amount of time he was in contact with English during the classes was not enough to learn the language well, as life outside the classroom continued to be in Arabic and French.

To practice the language and be forced to speak it daily, he decided to go abroad to study English in an English-speaking country. Issam was convinced that through this "full immersion" into the English language he would finally be able to improve his skills.

The linguistic ideology of language immersion assumes that staying in an environment where the target language is spoken is the most effective method to learn a language and pushes learners like Issam to become mobile. This ideology further includes the imagination of language tourism destinations as mostly monolingual places, where all conversation opportunities will be in the target language (Schedel, 2022)—an idea that proved not to be true, as we will see later.

Issam chose Malta out of other typical Western destinations for English language tourism for several reasons. The USA, Canada, Australia, or New Zealand were unreachable for him due to complicated visa procedures and high costs. Europe was close and he was eligible for a visa that allowed him not only to enter one nation-state but also to travel across the EU Schengen Area. Having watched lots of YouTube videos of young travelers, he was craving to explore the world himself. Also, according to his internet research, Malta—where most of the people were English native speakers, as he emphasized—was the cheapest European destination. Furthermore, Malta was known as a tax haven and seemed therefore to also be a great location to start a business.

Issam picked a business-oriented language school that offered traditional English courses as well as business English courses and MBA programs. He contacted an Arabic-speaking promoter of this school to get an invitation letter that he needed for his visa application. He planned to stay for eight months and had to pay the price of 3000 € for the language classes for his complete stay in advance to get this letter. This price was a tremendous amount for someone from his country of origin, where the average monthly salary is around 200–300 € (as Issam stated and online salary surveys such as Paylab.com confirm). With his parents having their own company and earning more than the average, he received the necessary financial support to get the visa and to pay the school fee. Nevertheless, this language holiday was a real investment, which also resonated in the interview as he repeated the word "investment" many times. With this "investment," he also took a great risk. If his visa application or its extension (the language tourist visa is always only issued for a period of 90 days and needs to be extended subsequently) were refused, the school would not have refunded the cost of the classes. Making investments into one's own personal development and taking risks is, however, part of the mindset of the neoliberal subject (Bröckling, 2007; De Costa et al., 2016).

Once he arrived in Malta, Issam had to navigate different challenges: for instance, he tried hard to find a convenient accommodation, but no one would give him a rent contract for the length of his stay, which he needed for his visa. In Malta, many apartments are rented informally and/or their owners prefer renting out pricy rooms to tourists for a short term—especially in the high season—instead of to long-period tenants. In the end, the owner of the hostel where he stayed during his room search gave him a contract for the length of his stay, but Issam had to sleep in a 12-person dormitory and paid the same price (450 €) for a bunkbed as for a single room. He also discovered he could not open a bank account and his parents had to pay high fees when they wanted to send him money.

These were, however, not the only issues that Issam had to overcome during his first weeks in this foreign country. In fact, Issam was struggling a lot since his English was at a very low level, and he continuously encountered language barriers. To get information, he drew on the help of other Arabic- or French-speakers he met in the hostel and used them as interpreters in urgent matters. Besides the language classes that he visited 3 h/day at the language school, he tried to "integrate"[4] into English—as he called it—also in his free time. He watched videos or read books in English and spent much time in the hostel's common area where he met lots of international people, who he first listened to and later tried to engage in conversations once he was more confident. After two or three months, he began to understand the people around him—even "native speakers" as he said—and to make himself understood. He found himself making many international friends, with whom he toured the island and enjoyed Malta's vibrant nightlife.

[4] He presumably meant "immersion" but did not know the exact word. However, I found it curious that he called it "integration" as his immersion into English encompassed social integration, namely, him trying to become a member belonging to an imagined community of international English-speakers.

When I interviewed Issam for the first time in April 2018, he told me that he was very happy about his learning progress over the past four months and that he was learning so much faster than his Arabic colleagues from the language school as they were hanging out with other Arabic-speakers while he was actively seeking out opportunities to practice English and avoiding other Arabic- or French-speakers. For him, the hostel with its international staff and guests was therefore the ideal place to learn English. In the interview, he described himself as a successful learner with agency over his learning outcomes, which aligns with the common discourse on immersion that responsibilizes learners for their achievements. However, this self-responsibilization is also part of a neoliberal rationality that transfers any responsibility of success to the entrepreneurial subjects themselves (see also Schedel, 2022).

When speaking to the promoter of the language school prior to coming to Malta, the promoter had (wrongly) told Issam that he could work in Malta to finance his stay. Issam thus planned to work part-time besides the classes to save some money as seed capital for his business. He managed to find a job, just to find out that he was not entitled to work with his visa.[5] Moreover, as his visa required 80% attendance, the school threatened to kick him out and send him back home, since he had skipped classes to go job hunting. Issam felt betrayed by the school and his motivation to study there decreased. In the interview, he called the language school's business a "bezness."[6] However, since he had already paid for the whole stay and did not want to lose his visa, he continued attending the classes.

[5] In 2018, Malta introduced a new law that allows for third country nationals who hold a language student visa and stay for longer than 90 days to work a maximum of 20 h/week starting from the 91st day. However, they have to continue to attend classes with a minimum of 15 h/week. When Issam arrived in Malta, this was not possible yet.

[6] This term, a combination of the English word "business," and the French word *baiser* (to fuck), is used in Maghreb countries to describe a rip-off—literally a business where you get fucked.

Desire for Freedom and Emancipation: Escaping or Being Trapped as a Voluntourist in Europe

After eight months in Malta, Issam's visa came to an end, but Issam did not want to go back home where he had felt depressed. He wanted to stay in Europe at any price. However, his longing for Europe seemed now to be sparked by a strong desire for freedom and emancipation. In an interview, he explained this new motivation retrospectively:

> Issam: Moi, déjà après mon expérience à Malte, tu vois, ma ma vision a commencé à changer. Moi, je commençais à changer. Et du coup, pour moi, revenir [au pays d'origine] c'était (.) c'était revenir en l'enfer quoi. Je pouvais pas avec la mentalité, [respire] avec comment les gens pensent. Donc, euh, pour moi c'était trop difficile d'y retourner. (...) Parce que, [au pays d'origine], la mentalité, elle est un petit peu, un petit peu enfermée, tu vois. Genre, t'as la mentalité des gens, t'as beaucoup de choses dans le pays qui sont pas beaux et t'as aussi la famille. La famille qui te font de la pression, qui sont toujours collés à toi, qui te laissent pas genre expérimenter la vie. Donc là-bas, quand je sors le soir, t'as tes parents qui commencent à t'appeler, tu vois. Il faut toujours que tu restes à côté de tes parents. Et ça, je voulais plus à partir d'un certain moment. Tu vois ? J'avais besoin de liberté. De toute façon, moi je suis ici pour ma liberté, tu vois, liberté d'expression, liberté de voyager, liberté de faire ce que je veux.
>
> ***My translation:***
> Issam: After my experience in Malta, you know, my, my vision has started to change. I began to change. Thus, for me going back to [his country of origin] that would (.) that would have been like going to hell. I couldn't bear with the mentality, [inhales] how people think. So, uhm, for me, going back would have been too difficult (...) Because, in [country of origin], the mentality is a little bit, a little bit closed-minded, you know. There is the peoples' mentality, there are many things in this country that aren't nice, and there is also the family. The family who puts pressure on you, who is always around you, who doesn't let you experience life. So, back home, when I go out at night, my parents will start calling me, you know. You must always stay close to your parents. And all these things, I didn't want this anymore at a certain moment, you know? I needed some freedom. After all, I'm here for my freedom, you know, freedom of expression, freedom to travel, freedom to do anything I want.

In this interview excerpt, Issam described his trip to Malta as a life-changing transition and Europe as a place where he could escape the surveillance of his family. Later in the interview, he further added he was also trying to break free from religious restrictions of Islam that he was experiencing in his home country, which did not allow him to have sex, to drink alcohol, or to go clubbing—all the things that he could enjoy in Malta. Before his trip to Malta, Issam had never left his hometown or lived on his own. His trip to Europe and the "freedoms" he experienced therefore have to be understood as a coming-of-age process. His strengthened desire for Europe and for the European values of freedom of expression, of travel, of religion, etc., turned his short language trip into a long-term project of emancipation, namely, his attempt to break away from the parental home and from a culture he perceived as too conservative (see also Piller & Takahashi, 2006, as well as Marty Crettenand, 2016, for similar emancipation projects of language learners on the move). His objective was now to gain autonomy and make his own life choices.

This desire for personal freedom and autonomy aligns well with the appeal of freedom of neoliberal discourses that constitute individuals as active, self-steering, and responsibility-taking agents who control their own lives. Neoliberal discourses thus invite individuals to break free from whatever holds them back, while at the same time also blaming them for their own suffering if they remain passive (Pyysiäinen et al., 2017).

To realize his dream of a free life in Europe, Issam needed to get a new visa or a residence permit. The Maltese language school advised him to do an MBA at their facility, which would again have cost a lot of money. Since he was completely dissatisfied with the school, he declined. He had an uncle living in Northern Italy who told him that he could stay with him and he would help him get a residence permit. So, he went to Italy and tried to change his language student visa into a student visa to do a second master's degree, which would have also allowed him to work in Italy. However, he was told that he had to go back to his country of origin to apply for the visa, which would not only have taken a long time, but his chances were also low to get the visa since he had already completed his master studies. Issam's desire for a life in Europe was, however, so strong that going back to his home country and risking remaining there was not an option for him. In the end, he decided to stay "illegally" in Europe after his language student visa expired.

Since staying with his uncle did not go well, Issam had to leave his uncle's flat. To avoid living on the street or having to do illicit work—as he saw many compatriots in Italy doing—Issam chose to do voluntourism as a survival strategy as it guaranteed him a bed, at times also a meal or a small renumeration when working overtime. His parents supported him financially as he did not earn enough—if anything—to make a living from it. What followed was an odyssey through the EU Schengen Area over a period of four years, during which Issam engaged in various forms of volunteer labor advertised to voluntourists. Meanwhile, he tried to get a residence permit with the help of different advocates—first in Italy, then in Spain. Issam also had a brother who was living in France whom he visited once, but he never even thought about trying to make a living there—although he was fluent in French—as he had a negative image of France and was convinced that he had no chance of getting a residence permit there.

Issam found his volunteer jobs on voluntourism websites such as HelpX, Worldpackers, or Workaway. The jobs he was hired for ranged from being a babysitter in a French-speaking Tunisian family in Italy and in a Spanish-speaking family with French skills in Spain to an English teacher in a private school in Italy to a housekeeper or a receptionist in different hostels with international clientele. Oftentimes, it was thanks to his French or English skills that he got these jobs, either because he had one of these languages in common with his hosts, which facilitated communication with them—especially since in the beginning he did not speak a word of the national languages Italian or Spanish. Most of the times, his English skills were, however, a key requirement for the volunteer jobs (e.g., for the teaching as well as for the receptionist jobs), while Italian or Spanish skills were not required. Issam saw himself evolving from being categorized as a beginner and later an intermediate learner of English in Malta to a competent speaker of English in Italy and Spain, where a workforce with advanced English skills was scarce.

All the workplaces where he volunteered turned out to be very multilingual contexts that allowed Issam to not only use his multilingual repertoire but also learn Italian and Spanish to differing degrees. While he was still staying with his uncle in Italy, he attended an Italian language class as he was convinced that he would need to speak proper Italian to be able to stay, study, and work in Italy. However, due to his visa problems, he was unable to focus on the class. Later, during his volunteer job at the school, he was able to pick up some Italian words from the children he did English activities with (see also Sinervo, this volume). While working in the hostels in Italy, it became, however, nearly impossible to further improve his Italian skills as these hostels were truly international spaces and lacked Italian clients or co-workers. He could, however, practice and improve his English at work. In Spain, he encountered the same problem of the absence of Spanish clients or co-workers; however, he met other travelers from South America from whom he could learn some basics in Spanish. In an interview, he told me he had been able to acquire a level of Spanish where he could communicate quite well and that he was studying on his own, but he needed to do a language class to get the grammar basics right. His experience of learning a language through immersion while working abroad contradicts the common assumptions of the "abroad" as a monolingual place where the national language is spoken everywhere and as the ideal opportunity to learn a language the "natural" way.

Besides being a means of survival, the volunteer jobs also served Issam as a means of social integration (Ambrosini & Artero, 2022). Instead of renting a room and being on his own, voluntourism allowed him to work and to socialize with other young people in the hostels or the families with whom he stayed. However, the working and housing conditions as a volunteer were at times extremely precarious and he was exploited. For instance, some of his hosts forced him to work in the event of illness, to do unpaid overtime, provided precarious housing such as sleeping in a school or in a dormitory with many other people, and treated him without respect. One host used his illegalized status as an excuse to let him do only nightshifts and made him work over 50 h/week. Furthermore, as voluntourists are often provided with an accommodation that is in their workplace, the boundaries between work and leisure became blurred (Rice, 2010). Issam experienced this several times: for example, while volunteering in a family where he could not clearly distinguish between work and free time, as his job as a babysitter comprised daily family care work (see also McDaid & Sunyol, this volume), but also while volunteering and staying in hostels (see also Schedel, 2022). He suffered from sleep problems, a poor diet, and unhygienic conditions, and he was severely ill several times. Since he had no health insurance, seeing a doctor and buying medicine was always a particularly expensive venture (see also McDaid & Sunyol, this volume).

While I observed regular voluntourists quitting their volunteer jobs when experiencing bad working or housing conditions as well as disrespectful treatment, for Issam it was not that easy. He was dependent on these volunteer jobs and the accommodation provided. If he did not want to end up on the street, he first had to find another volunteer job, which became difficult during the COVID-19 pandemic and during the lockdowns even impossible. Consequently, he had to rent a room and wait. In case of mistreatment or exploitation, Issam could furthermore not afford to report it and was left without the protections of social legislation due to his own illegalized status.

Despite these precarious living and working conditions in "illegality," Issam did not think of going back to his home country. He was convinced that sooner or later he would get his residence permit and that all the hardships were worth it. Toward the end of our last interview, he mentioned that if he had stayed in his home country, he would certainly already be married and have a family, but this was not for him. Therefore, instead of only seeing the problematic sides, he preferred enjoying the positive things that his life as a voluntourist in Europe could offer him (see also Choi, this volume): travelling—for which he had developed a great passion—seeing beautiful places, making international friends, and going out and spending his leisure time as he wanted without anyone watching or disciplining him. Yet, he acknowledged that he was not totally free yet, either. Further, he was not independent from his parents, as they were still financing him. When I asked about his dreams of becoming an entrepreneur, he sounded resigned. He had abandoned all his original plans and the only thing that still mattered to him after all this time was to obtain his residence permit.

Young people's international mobility such as voluntourism is often described as a rite of passage (Gabowski et al., 2017) that comes along "with a number of youth life transitions—from education to work, from unemployment to employment, and, more widely, from 'youth' to 'adulthood'" (King et al., 2016, n.p.). The geographical mobility of volunteers is hence associated with personal growth and imagined to allow for economic/social mobility. The idea of a transition also refers to a time limitation, with a clear starting and end point. From the time perspective, tourism is also usually understood as a temporary mobility and migration as a permanent one, whereas it is not clear how temporariness or permanency are defined (see Williams & Hall, 2000, for a critique). However, Issam's volunteer experience spanned not just over some weeks or months but over several years. Shanthi Robertson (2013) describes this long period of temporariness and uncertainty from entering a nation-state with a student visa to gaining a residence permit with the metaphor of a "long tunnel." For a long time, Issam could not see the light at the end of this tunnel. While on the one hand, he was constantly on the move, changing places and volunteer jobs, his mobility was on the other hand "bounded" (Gutekunst et al., 2016). Due to his illegalized status in Europe, he could not move where he wanted (once he was even arrested while crossing the border to Spain but was set free again after a couple of hours), and he was also facing immobility as there was no improvement of his precarious situation. He was stuck in this "passage" and the liminal "rite" had become his life(style) (Cohen, 2011; Nakano, 2000).

Eventually, one day, this temporariness came to an end. In October 2022, Issam informed me that, with the help of his advocate, he had succeeded in getting a residence permit for Spain. His message seemed relieved, full of hope and new dreams of finding a paid job, of getting ahead in his life, and of finally being able to enjoy his hard-won freedom in Europe.

Conclusion: Toward a More Comprehensive Understanding of (Inequalities in) Voluntourism

This study explored how an illegalized migrant from Northwest Africa used voluntourism as a survival strategy to bridge the long period of uncertainty trying to get a residence permit for any European country. In fact, volunteer work allowed for ensuring basic needs (shelter, a meal, occasionally a small renumeration) and social integration. The chapter also investigated the (language) desires and ideologies that made our protagonist, Issam, leave a safe job in his home country, look for different language learning opportunities abroad, and endure the hardships he encountered as a voluntourist.

The analysis of Issam's narrated trajectory reveals a complex bundle of desires for entrepreneurship, for Europe and, more broadly, the West, for English, and to a lesser extent also for Spanish and Italian, as well as the desire of a young adult to emancipate himself from his parental home and to escape a religious community that he perceived as too conservative. These desires were strengthened by different language ideologies, such as the ideology of English as a global language and the language of business, the ideology of language immersion as the most effective language learning method, and the ideology of linguistic nationalism that entails the belief that a good command of skills in the national language are necessary to have access to a nation-state's local job market. His journey taught him, however, that these common assumptions about language(s) are too simplistic and that the situation is oftentimes more complex.

The analysis further shows that Issam's language learning and voluntourism practices were overall governed by a neoliberal ideology that constitutes individuals as entrepreneurial selves who must invest in their own skilling (e.g., learning English) to self-optimize the value of the own persona on a global market. The entrepreneurial mindset of these neoliberal subjects also implies working hard, being eager and persistent, taking risks, and searching for freedom, which goes hand in hand with a greater self-responsibilization for the outcomes of one's choices and actions (Bröckling, 2007).

While neoliberal governance often works through the appeal of freedom (Pyysiäinen et al., 2017), the realities of neoliberal subjects searching for personal freedom and self-realization may, as we have seen, look very different and lead to deep precarity and new dependencies instead. In fact, while Issam's dream of making a living in Europe was bolstered by the desire for upward social mobility and emancipation from his parental home and cultural constraints in his country of origin, he ended up submitting himself to extremely precarious living and working conditions as a volunteer and was still financially dependent on his parents.

Issam's story not only unsettles the usual depiction of voluntourism as a North-South mobility of White "Barbie Saviors" (Wearing et al., 2018), but it also allows for unpacking inequalities among voluntourists depending on their origins and on their (il)legalized status (see also Schedel, 2022, on the hierarchization of voluntourists according to their language skills). His story further highlights the fuzziness of supposedly dichotomic categories such as "tourist/migrant (worker)," "leisure/work," and "im-/mobility," which, in Issam's case, became a question of presentation, of interpretation, of being fluid and intersecting (see also Rice, 2010, and Salazar, 2022).

By giving a voice to a type of voluntourist that has been overlooked in the literature, this study has extended our understanding of motives for, experiences, and social consequences of voluntourism practices today. As the detailed exploration of his case has shown, some common assumptions about voluntourism need to be revised and further research on voluntourism "off-the-beaten-track" is needed. To reach a more comprehensive understanding of contemporary voluntourism practices, a focus on marginalized types of transnational volunteer workers who, at first glance, do not fit the common definition of voluntourists might constitute a rich pool for investigation.

Acknowledgments I thank "Issam" for trusting me, for dedicating his time, and for sharing his story. I am grateful to Cori Jakubiak for her valuable feedback on previous versions of the manuscript, to Jennifer Raab for proofreading the manuscript, as well as to Katy Highet and Beatriz Lorente for reflecting with me on how to write about "illegality."

Research Funding This work was supported by the Office for Gender Equality of the University of Bonn (Maria von Linden Program). During the fieldwork, I was a guest researcher at the University of Malta.

References

Ambrosini, M., & Artero, M. (2022). Immigrant volunteering: A form of citizenship from below. *VOLUNTAS: International Journal of Voluntary and Nonprofit Organizations, 34*, 252–262. https://doi.org/10.1007/s11266-022-00454-x

Baillie Smith, M., & Laurie, N. (2011). International volunteering and development: Global citizenship and neoliberal professionalisation today. *Transactions of the Institute of British Geographers, 36*(4), 545–559. http://www.jstor.org/stable/23020828

Baillie Smith, M., Laurie, N., & Griffiths, M. (2018). South–South volunteering and development. *The Geographical Journal, 184*(2), 158–168. https://doi.org/10.1111/geoj.12243

Baillie Smith, M., Fadel, B., O'Loghlen, A., & Hazeldine, S. (2022). Volunteering hierarchies in the global South: Remuneration and livelihoods. *VOLUNTAS: International Journal of Voluntary and Nonprofit Organizations, 33*, 93–106. https://doi.org/10.1007/s11266-020-00254-1

Brennan, C. (2014). Precarious working tourists: Working holiday makers in Australia. [Doctoral dissertation, University of Tampere]. Trepo Repository. https://trepo.tuni.fi/bitstream/handle/10024/95218/978-951-44-9438-3.pdf?sequence=1&isAllowed=y

Brennan, C. (2018). Northern lights instead of workers' rights: Volunteer working tourists in Finnish Lapland. *Nordic Journal of Working Life Studies, 8*(2), 43–61. https://doi.org/10.18291/njwls.v8i2.106154

Bröckling, U. (2007). *Das unternehmerische Selbst: Soziologie einer Subjektivierungsform*. Suhrkamp.

Busch, B. (2017). Expanding the notion of the linguistic repertoire: On the concept of Spracherleben—The lived experience of language. *Applied Linguistics, 38*(3), 340–358. https://doi.org/10.1093/applin/amv030

Butcher, J., & Smith, P. (2015). *Volunteer tourism: The lifestyle politics of international development*. Routledge. https://doi.org/10.4324/9781315796376

Cohen, S. A. (2011). Lifestyle travellers: Backpacking as a way of life. *Annals of Tourism Research, 38*(4), 1535–1555. https://doi.org/10.1016/j.annals.2011.02.002

Coghlan, A., & Fennell, D. (2009). Myth or substance: An examination of altruism as the basis of volunteer tourism. *Annals of Leisure Research, 12*(3–4), 377–402. https://doi.org/10.1080/11745398.2009.9686830

De Costa, P., Park, J., & Wee, L. (2016). Language learning as linguistic entrepreneurship: Implications for language education. *The Asia-Pacific Education Researcher, 25*, 695–702. https://doi.org/10.1007/s40299-016-0302-5

Dlaske, K. (2016). Shaping subjects of globalisation: At the intersection of voluntourism and the new economy. *Multilingua, 35*(4), 415–440. https://doi.org/10.1515/multi-2015-0002

Doerr, N. M., & Taieb, H. D. (Eds.). (2020). *The romance of crossing borders: Studying and volunteering abroad*. Berghahn Books.

Flubacher, M.-C. (2022). Ethnography as a speaker-centred approach? Methodological reflections. In J. Purkarthofer & M.-C. Flubacher (Eds.), *Speaking subjects in multilingualism research: Biographical and speaker-centred approaches* (pp. 51–61). Multilingual Matters.

Flubacher, M.-C., & Purkarthofer, J. (2022). Speaking subjects in multilingualism research: Biographical and speaker-centred approaches. In J. Purkarthofer & M.-C. Flubacher (Eds.), *Speaking subjects in multilingualism research: Biographical and speaker-centred approaches* (pp. 3–20). Multilingual Matters.

Gabowski, S., Wearing, S., Lyons, K., Tarrant, M., & Landon, A. (2017). A rite of passage? Exploring youth transformation and global citizenry in the study abroad experience. *Tourism Recreation Research, 42*(2), 139–149. https://doi.org/10.1080/02508281.2017.1292177

Ghazarian, P. G. (2021). Youth as temporary workers abroad: The experiences of Australia, Canada and New Zealand. In D. Cairns (Ed.), *The Palgrave handbook of youth mobility and educational migration* (pp. 323–332). Palgrave Macmillan. https://doi.org/10.1007/978-3-030-64235-8_29

Gomberg-Muñoz, R. (2016). Criminalized workers: Introduction to special issue on migrant labor and mass deportation. *Anthropology of Work Review, 37*, 3–10. https://doi.org/10.1111/awr.12080

Gutekunst, M., Hackl, A., Leoncini, S., Schwarz, J., & Götz, I. (2016). *Bounded mobilities: Ethnographic perspectives on social hierarchies and global inequalities.* Transcript. https://doi.org/10.1515/9783839431238

Hamlin, R. (2021). *Crossing: How we label and react to people on the move.* Stanford University Press.

Heller, M. (2002). *Éléments d'une sociolinguistique critique.* Didier.

Jang, I. C. (2017). *Consuming global language and culture: South Korean youth in English study abroad.* [Doctoral dissertation, University of Toronto]. Tspace Repository. https://tspace.library.utoronto.ca/bitstream/1807/79006/3/Jang_In_Chull_201706_PhD_thesis.pdf.

King, R., Lulle, A., Morosanu, L., & Williams, A. (2016). *International youth mobility and life transitions in Europe: Questions, definitions, typologies and theoretical approaches.* Report. University of Sussex. https://hdl.handle.net/10779/uos.23430644.v1

Kinginger, C. (2004). Alice doesn't live here anymore: Foreign language learning and identity reconstruction. In A. Pavlenko & A. Blackledge (Eds.), *Negotiation of identities in multilingual contexts* (pp. 219–242). Multilingual Matters. https://doi.org/10.21832/9781853596483-010

Kinginger, C. (2009). *Language learning and study abroad: A critical reading of research.* Palgrave Macmillan. https://doi.org/10.1057/9780230240766

Kramsch, C. (2006). Desire in language: The neglected dimension of language learning. In S. Duxa, A. Hu, & B. Schmenk (Eds.), *Grenzen überschreiten. Menschen, Sprachen, Kulturen. Festschrift für Inge Christine Schwerdtfeger zum 60. Geburtstag* (pp. 209–224). Gunter Narr.

Lyons, K., Hanley, J., Wearing, S., & Neil, J. (2012). Gap year volunteer tourism: Myths of global citizenship. *Annals of Tourism Research, 39*(1), 361–378. https://doi.org/10.1016/j.annals.2011.04.016

Marty Crettenand, S. (2016). Le rôle de la langue dans le projet de migration de jeunes femmes suisses alémaniques vers la Suisse romande. In M. Matthey & A. Millet (Eds.), *Hétérogénéité et changement: Perspectives sociolinguistiques. Actes du 2ème congrès du réseau francophone de sociolinguistique, Grenoble, 10–11 juin 2015* (pp. 109–116). EME Editions.

McGloin, C., & Georgeou, N. (2016). Looks good on your CV: The sociology of voluntourism recruitment in higher education. *Journal of Sociology, 52*(2), 403–417. https://doi.org/10.1177/1440783314562416

Meyer Pitton, L., & Schedel, L. S. (2022). The (hidden) politics of language choice in research on multilingualism: Moments of (dis)empowerment. In P. Holmes, J. Reynolds, & S. Ganassin (Eds.), *The politics of researching multilingually* (pp. 307–326). Multilingual Matters. https://doi.org/10.21832/9781800410152-019

Mostafanezhad, M. (2014). *Volunteer tourism: Popular humanitarianism in neoliberal times*. Routledge. https://doi.org/10.4324/9781315547886

Motha, S., & Lin, A. (2014). "Non-coercive rearrangements": Theorizing desire in TESOL. *TESOL Quarterly, 48*(2), 331–359. https://doi.org/10.1002/tesq.126

Nakano, L. Y. (2000). Volunteering as a lifestyle choice: Negotiating self-identities in Japan. *Ethnology, 39*(2), 93–107. https://doi.org/10.2307/3773837

Park, J. S.-Y. (2010). Naturalization of competence and the neoliberal subject: Success stories of English language learning in the Korean conservative press. *Journal of Linguistic Anthropology, 20*(1), 22–38. https://doi.org/10.1111/j.1548-1395.2010.01046.x

Park, J. S.-Y. (2011). The promise of English: Linguistic capital and the neoliberal worker in the South Korean job market. *International Journal of Bilingual Education and Bilingualism, 14*(4), 443–455. https://doi.org/10.1080/13670050.2011.573067

Paylab Salary Survey. (n.d.). Retrieved November 30, 2022, from www.paylab.com

Piller, I., & Takahashi, K. (2006). A passion for English: Desire and the language market. In A. Pavlenko (Ed.), *Bilingual minds: Emotional experience, expression and representation* (pp. 59–83). Multilingual Matters. https://doi.org/10.21832/9781853598746-005

Pyysiäinen, J., Halpin, D., & Guilfoyle, A. (2017). Neoliberal governance and 'responsibilization' of agents: Reassessing the mechanisms of responsibility-shift in neoliberal discursive environments. *Distinktion. Journal of Social Theory, 18*(2), 215–235. https://doi.org/10.1080/1600910X.2017.1331858

Reilly, A. (2015). Low-cost labour or cultural exchange? Reforming the Working Holiday visa programme. *The Economic and Labour Relations Review, 26*(3), 474–489. https://doi.org/10.1177/1035304615598160

Remers, I. (2022). Voluntourism in Sub-Saharan Africa is expiation by the West, but only creates further dependency on the West. *Glocality, 5*(1–2), 1–17. https://doi.org/10.5334/glo.50

Rice, K. (2010). 'Working on holiday': Relationships between tourism and work among young Canadians in Edinburgh. *Anthropology in Action, 17*(1), 30–40. https://doi.org/10.3167/aia.2010.170104

Robertson, S. (2013). Encountering the residence regime. In *Transnational student-migrants and the state: The education-migraton nexus* (pp. 84–99). Palgrave Macmillan. https://doi.org/10.1057/9781137267085_5

Ruhs, M., & Anderson, B. (2010). Semi-compliance and illegality in migrant labour markets: An analysis of migrants, employers, and the state in the UK. *Population, Space, and Place, 16*(3), 195–211. https://doi.org/10.1002/psp.588

Salazar, N. B. (2022). Labour migration and tourism mobilities: Time to bring sustainability into the debate. *Tourism Geographies, 24*(1), 141–151. https://doi.org/10.1080/14616688.2020.1801827

Schedel, L. S. (2022). The price of immersion: Language learners as a cheap workforce in Malta's voluntourism industry. *Multilingua, 41*(2), 181–200. https://doi.org/10.1515/multi-2020-0178

Stainton, H. (2016). A segmented volunteer tourism industry. *Annals of Tourism Research, 61*(C), 256–258. https://doi.org/10.1016/j.annals.2016.09.011

Takahashi, K. (2013). Language learning, gender and desire: Japanese women on the move. *Multilingual Matters.* https://doi.org/10.21832/9781847698551

Tomazos, K., & Butler, R. (2009). Volunteer tourism: Working on holiday or playing at work? *Tourismos: An International Multidisciplinary Journal of Tourism, 4*(4) 331–349. http://www.chios.aegean.gr/tourism/vol4no4.pdf

Tiessen, R. (2018). *Learning and volunteering abroad for development: Unpacking host organization and volunteer rationales.* Routledge. https://doi.org/10.4324/9781351709415

Williams, A. M., & Hall, C. M. (2000). Tourism and migration: New relationships between production and consumption. *Tourism Geographies, 2*(1), 5–27. https://doi.org/10.1080/146166800363420

Vrasti, W., & Montsion, J. M. (2014). No good deed goes unrewarded: The values/virtues of transnational volunteerism in neoliberal capital. *Global Society, 28*(3), 336–355. https://doi.org/10.1080/13600826.2014.900738

Wearing, S. L. (2001). *Volunteer tourism: Experiences that make a difference.* CABI. https://doi.org/10.1079/9780851995335.0000

Wearing, S., Mostafanezhad, M., Nguyen, N., Nguyen, T. H. T., & McDonald, M. (2018). "Poor children on Tinder" and their Barbie Saviors: Towards a feminist political economy of volunteer tourism. *Leisure Studies, 37*(5), 500–515. https://doi.org/10.1080/02614367.2018.1504979

7

Institutionalized Volunteerism in Language Tourism: Volunteer Internship Programs for South Korean Young Adults Studying English in Toronto

In Chull Jang

Introduction

Teaching English as a foreign language (TEFL) has been a major area of interest in the study of voluntourism (Jakubiak, 2016, 2020; Stainton, 2019). While native English-speaking tourists are generally not certified language teachers, they often volunteer in TEFL activities for local students. Similarly, during the course of language tourism (Iglesias, 2016)—a type of educational travel for language learning for a relatively short period of staying abroad—a number of English language learners seek out volunteering opportunities to acquire and use the target language in natural settings. The primary motivation for engaging in volunteerism during language study abroad programs is what Doerr (2015) calls the discourse of "immersion": "through experiencing mundane daily life

I. C. Jang (✉)
Seoul National University, Seoul, Republic of Korea
e-mail: icjang@snu.ac.kr

L. S. Schedel, C. Jakubiak (eds.), *Voluntourism and Language Learning/Teaching*, Palgrave Advances in Language and Linguistics,
https://doi.org/10.1007/978-3-031-40813-7_7

while immersed in the host city, students are expected to develop the sensibility, attitudes, skills and knowledge that make up (…) global competence" (p. 369). This discourse emphasizes that students should immerse themselves in the local community, interact and communicate with locals and experience authentic cultures to accomplish a "successful" education abroad (Doerr, 2018). Thus, it frequently leads students and educators to see learning contexts outside of the classrooms as more advantageous and effective for achieving language learning goals. Specifically, in the discourse of language tourism, immersion into local communities is further associated with native speakerism, as these places are imagined as ones where language learners can meet "native" speakers (Han, 2014; Park, 2010a). In this sense, similar to TEFL voluntourism, volunteerism in English language learning is based on the monoglossic ideology that native English speakers are ideal and desirable language models. However, English learners as volunteers do not offer immersion experience to locals in need; rather, they seek immersion experiences with locals by engaging in volunteer work.

In fact, language tourism has presented volunteerism as one of the most effective language learning programs for studying language abroad. Volunteerism is said to provide students with a variety of language learning opportunities and experiences, and new language learning programs have been developed to meet students' changing (linguistic) demands. In recent years, study abroad agencies and private language schools have incorporated a variety of volunteer activities into their curricular or extracurricular language learning programs, including language exchange, participation opportunities in local altruistic organizations, and internships in local businesses. Consequently, volunteerism has been institutionalized or commercialized in the context of language tourism (Stainton, 2019).

Using Biao Xiang and Johan Lindquist's (2014) concept of migration infrastructure, this study examines the institutionalization of volunteer opportunities for international English language students who are seeking immersion experiences. According to Xiang and Lindquist (2014), migration should be "conceptualized through a focus on infrastructure rather than on state policies, the labor market, or migrant social networks alone" (p. S122). That is, while each stakeholder, institution, or technology operates according to its own logic of migration, they are materially and discursively interconnected in order to promote a specific type of global migration. Moreover, Xiang (2012) argues that in a complex system of governance for migration, brokers play a crucial role not only in recruiting and arranging the itineraries of prospective migrants, but also in producing and expanding the market of migration for autonomous agents (cf. Ambrosini, 2017; Lorente, 2018). Although language tourism is different from other forms of migration in terms of patterns and motivations, study abroad agencies and language schools in the language tourism industry play a brokering role by promoting volunteerism for language learning, mediating it to study abroad students, and maximizing the linguistic and other material benefits of students' participation—that is, institutionalizing volunteerism. Thus, this study analyzes volunteer internship programs catered to international English language students to illustrate the discourses and practices of institutionalized volunteerism as well as the consequences for the infrastructure of language tourism as a form of short-term migration.

The case of language tourism that this study will analyze is what is called *yeongeo eohakyeonsu* (literally, training English abroad) among South Korean youth. This type of language tourism is considered a self-development project, as young adults are motivated to improve their English communication skills by spending six to twelve months studying in an English-speaking country to increase their employability in the globalizing South Korean job market (Jang, 2015). This transnational youth mobility differs from other study abroad programs supported by universities or colleges. Korean students enroll in private ESL (English as a second language) institutes or language centers affiliated with colleges or universities in an English-speaking country as an individual and self-funded project to improve their English. When Korean students consider studying English abroad, they first consult a local study abroad agency called *yuhakwon* for information on destinations, language programs, length of study, and costs. When they finally decide to pursue this language training abroad, they sign a contract with an agent. Thus, the language education industry, which includes private English schools and study abroad agencies, plays an important role in creating and managing students' learning plans.

Drawing on ethnographic fieldwork with South Korean students undertaking *yeongeo eohakyeonsu* in Toronto, Canada, this study will demonstrate (1) how intermediaries, such as study abroad agencies and private language schools, present internships as volunteer opportunities for international students and (2) how students perceive volunteer internships as linguistic and cultural immersion experiences. The study reveals that even though intermediaries play a crucial role in institutionalizing and commercializing this system of language tourism, volunteer internships are interconnected with students' desires, labor market demands, the host government's immigration policy, and the language industry's market logics. At the same time, this study shows that students demystify the promise of volunteer internships by being aware—often through their peers—of these internships' limitations in helping them obtain global experience and improve their English language proficiency.

Infrastructure of Language Tourism as Migration

To understand migration not from the position of "migrants who migrate, but rather constellations consisting of migrants and non-migrants, of human and non-human actors" (p. S124), Xiang and Lindquist (2014) propose the notion of "migration infrastructure—the systematically interlinked technologies, institutions, and actors that facilitate and condition mobility" (p. S124). They suggest that this concept can help uncover "what is hidden," "what is constructed from within," and "how migrants are moved by others" in the migration practice. Xiang (2012) further argues that in a networked system of governance for migration, brokers emerge as agents who play a crucial role in not only recruiting and arranging the itineraries of prospective migrants, but also producing and expanding the market of migration. Not only are migration programs and patterns becoming increasingly diverse and complex, but traditional institutions that control migration policies—such as central and local governments—are outsourcing their training and recruitment to private agencies. This process leads intermediaries in the infrastructure of migration to participate in "self-perpetuating and self-serving" the infrastructure itself. As key players in the infrastructure of migration, brokers not only recruit and dispatch migrants, but also rationalize the necessity of migration through marketable strategies, thereby self-serving or self-perpetuating their roles as well as the migration market. These intermediaries use various social discourses and ideologies to persuade their clients, and they find loopholes to circumvent institutional and bureaucratic constraints as this process leads to market expansion. Thus, the development of brokerage within the migration infrastructure contributes to an increased degree of institutionalization and commercialization of migration.

Montserrat Iglesias (2016) defines language tourism as "a tourist activity undertaken by those travellers (or educational tourists)" that "includes at least an overnight stay in a destination outside their usual place of residence for less than a year and for whom language learning is a primary or secondary part of their trip" (p. 31). This definition indicates that language tourism is a kind of short-term educational migration governed by the pursuit of foreign language and culture learning. Despite the growing significance of brokerage in migration, research on language study abroad in general and language tourism specifically has conducted little systematic analysis of practices and ideologies of "agents," "brokers," or "intermediaries." The majority of previous research has focused on the motivations and activities of students who study languages abroad and the challenges they face when negotiating their multiple identities (Block, 2007; Kinginger, 2009). This body of research tends to place migrating students at the center of analysis so as to figure out migration practices and ideologies instead of addressing the infrastructure of language tourism or the role of agents or brokers in this system.

In the language tourism industry, a number of institutions provide brokerage services by connecting students to a variety of educational services and developing new learning programs to meet their needs. The representative organization is a study abroad agency, or *yuhakwon*, that connects students with other study abroad institutions, such as private language schools and accommodations, and also delegates visa processes. In South Korea, and perhaps in other countries, language tourism is mediated not by tourism agencies which primarily cater to customers traveling for leisure but by study abroad agencies which serve those who are planning to study abroad in the formal education sector. When the language tourism market for English was expanding due to globalization in the early 1990s, study abroad agencies took the niche market, leading educational services for English language tourism. As agents in these study abroad agencies are usually those who have already experienced language tourism, they have detailed knowledge and strategies regarding successful language learning abroad to share with their clients. Thus, a study abroad agency is primarily a site where a number of educational resources, knowledge, and methods are commodified and consumed (Shin, 2016). Moreover, recently, private English language schools—particularly large, franchised institutions—have begun to operate their own departments for internships, volunteerism, and homestays.

The agency is a private, for-profit participant in the infrastructure of language tourism that connects educational providers and consumers in incoming and outgoing countries. However, in contrast to other types of migration, in language tourism, the relationship between the agency industry and the government regulating educational and migration policies is relatively loose, and government regulations on agency businesses are not stringent. The South Korean government has attempted to regulate educational migration for English learning, believing it to be an outflow of national wealth (cf. Nelson, 2006). However, since the majority of English study abroad is self-funded and individually pursued in the private education sector, it is difficult for the liberal democratic government to regulate it within a non-authoritarian and capitalist system (Park & Abelmann, 2004). Conversely, the receiving countries, such as Canada, have promoted international education for the prosperity of their knowledge-based economy. Thus, the primary mission of a study abroad agency is to maintain demand for studying abroad in the face of students' ever-changing educational needs. Therefore, the industry of study abroad agencies constantly mobilizes social and educational discourses within a system of governance of student mobility to create a niche market for English language learning programs (Shin, 2016).

Global Experience in Neoliberal South Korean Society

The pursuit of global experiences by South Korean youth is a result of the country's high youth unemployment rate and increased job insecurity. The economic crisis of 1997, also known as the IMF crisis, has been recognized as a pivotal event in the erosion of job security in South Korean society. While the country underwent a neoliberal restructuring of its economy—a primary condition for its bailout provided by the International Monetary Fund (IMF)—the government implemented neoliberal labor policies, such as the provision for worker layoffs and the introduction of various forms of temporary employment (Song, 2009, 2011). These actions increased the flexibility and precariousness of the labor market, giving South Koreans the impression that lifelong job security is exceedingly difficult in a neoliberal workplace.

Consequently, today's young adults searching for employment face the bifurcation of the labor market: low-paying, temporary, low-skilled positions, and well-paying, high-skilled, and networked positions (Chun & Han, 2015; Yoon, 2014). Between these two types of employment are public employments, such as public service workers and teachers in public education, which offer lower pay but higher job security and pension benefits. The majority of young job seekers in South Korea struggle to find professional or public employment (Choi, this volume; Chun & Han, 2015; Yoon 2014). However, only a small number of talents are recruited and hired by prestigious corporations, and competition for public employment is fierce. Thus, the majority of young job seekers fall into the precarious and flexible labor force.

Phillip Brown and his colleagues have made substantial contributions to the mechanism of educational investment and training in the "jobless" neoliberal and knowledge-based economy. They refer to this phenomenon as *positional competition* (Brown et al., 2003, 2011). They argue that in a competitive job market, job applicants with superior skills and qualifications are placed in better and higher positions than those without. In light of the high youth unemployment rate and the fierce competition for better jobs, the failure to acquire even a *single* item could render one less competitive within a pool of qualified job applicants. This mechanism serves as a powerful incentive for job seekers to continually engage in developing and acquiring the skills and qualifications that they believe corporations require their employees to possess. In addition, in a knowledge-based economy, employability is determined not only by credentials, but also by *soft* skills acquired through a variety of experiences (Urciuoli, 2008). This mechanism is known as the *economy of experience* (Brown et al., 2003; Yoon, 2014). Successful global talents are the group of job applicants who achieve a competitive position on the labor market by acquiring a variety of global experiences. Volunteering and internship programs are viewed as providing job seekers with valuable workplace culture, professional knowledge, and social relationship experience (Allan, 2019).

The self-development of South Korean students for employability is perceived to be governed by the *spec* discourse. Hae-Joang Cho (2015) defines the discourse as follows:

> The term *spec* (an abbreviation of the word *specifications*—the detailed list of features describing the various components of a consumer product) pertains to resume-building activities and to the salient desire to attain long-term, secure employment. Upon entering a prestigious college with a lucrative major, students continue studying for the English proficiency test, aim for a high grade point average (GPA), prepare for various contests and qualification tests, and participate in study-abroad and internship programs. (pp. 445–456, emphasis in original)

Within this discourse, all educational investments are viewed as resume-building endeavors, and job seekers are expected to be goal-oriented and instrumentally motivated. Among these activities are study abroad and internships, which are regarded as requiring more financial and material resources than other qualifications and experiences. However, under the discourse of employability, going abroad is thought to be an advantageous experience because it is believed to provide students with opportunities to learn and/or use a foreign language (particularly English as a global language), understand foreign cultures, and cultivate intercultural and cosmopolitan sensitivity (Yoon, 2014).

Unsurprisingly, global internships are regarded as more valuable than studying abroad because they may provide additional value in the form of work experience in a foreign business (Jang, 2015).

The pursuit of global experience for employability is closely related to "English fever" (Park, 2009) in South Korean society. As English proficiency plays a crucial role in upward social mobility (e.g., university entrance and employment) in the globalized labor market, South Korean people have invested time and money into improving their English ability. However, what matters in their English investment for employment is the type of English they should acquire to make a distinction in the competition for a better position. In particular, as both employers and job seekers believe that a high score on a standardized English test does not guarantee actual English ability—especially on oral communication skills in English—they perceive living, working, and studying in an English-speaking country as evidence of a "real" command of English. This perception has served as another driver for South Korean youths' global experiences, including working holidays, language tourism, and global co-op programs.

The Study

This study utilized ethnographic fieldwork conducted among South Korean undergraduates undertaking *yeongeo eohakyeonsu* in Toronto, Canada. Its aim was to investigate the ways in which study abroad agencies institutionalize volunteer internships in the infrastructure of study abroad as a form of migration. The majority of fieldwork was conducted in 2013–2014 with 15 undergraduates and recent graduates from South Korea enrolled at Lingua City, a large private ESL school in downtown Toronto. I accessed this school as one of my colleagues, who was then teaching there, introduced me to it. She told me that because Lingua City provided a variety of English courses and programs, a great number of South Korean students selected it for their language learning. Once I obtained permission from Lingua City to conduct fieldwork, I followed students in and out of the school to observe and/or interview them. Using "following participants" as a methodological strategy (Marcus, 1995), I attended or participated in a variety of in-person and online social and educational events. I discovered that some of the participants who were dissatisfied with the classroom environments at the language school due to a lack of interactions with locals voluntarily pursued immersion opportunities through volunteer work in local communities (Jang, 2020). Moreover, study abroad agencies and language schools promoted volunteer experiences as a way for students to maximize language gains from their investment in English language learning. This was because internships, it was said, could offer opportunities to communicate with locals in authentic contexts, and study abroad agencies and language schools developed programs to facilitate successful volunteering. Therefore, I conducted additional fieldwork on a range of volunteer activities in which South Korean students participated, including visiting volunteering websites and attending volunteer orientations or actual events.

As a metropolitan city, Toronto offered a relatively large number of volunteer opportunities. It was in this context that study abroad agencies and English language schools could introduce such opportunities to students or that students could seek similar opportunities by themselves. Nonetheless, South Korean students typically discovered the limitations of volunteer work in improving their English learning and use. As I have explained elsewhere (Jang, 2020), the majority of volunteer opportunities for study abroad students consisted of seasonal and one-time local events and festivals that recruited volunteers for manual labor. Furthermore, even though South Korean students came to work with a group of local English-speaking volunteers, they often failed to communicate with them due to language barriers.

Still, as some study abroad agencies and language schools recognized volunteering as a means of experiential learning, they began to offer these opportunities to their clients. They also issued certificates of completion to those who fulfilled a certain amount of volunteer hours to encourage students to participate. This intervention of intermediaries in the local volunteering market reshaped the demography of volunteers; at some events, ESL students accounted for the majority of the volunteer labor force. These limitations of volunteer work led some South Korean students to seek out volunteer opportunities that would allow them to develop long-lasting relationships and engage in communication with locals. However, students' lack of a social network in Toronto made it difficult for them to search out and participate in such volunteer programs. In these situations, South Korean students viewed the volunteer internship programs presented and promoted by study abroad agents and private language schools as an alternative.

In focusing on the role of intermediaries in the development and promotion of volunteer internships in the English language tourism industry, I searched for and organized excerpts from interviews, observations, field notes, and documents collected during fieldwork that pertained to volunteer internships organized by language schools and study abroad agencies. The primary data set used in this project fell into three categories. First, the promotional materials of internship programs collected at study abroad fairs, on study abroad agency websites, and in the offices of study abroad agencies comprised the majority of data that revealed the intermediaries' discourses and practices. However, observations on the everyday practices of the study abroad agencies and interviews with agents were particularly limited due to their reluctance to allow me to conduct fieldwork or interviews out of concern for the disclosure of their business secrets and consulting strategies. Nevertheless, the promotional materials illustrated how study abroad agencies presented volunteer internship programs and the expectations these internships were believed to foster.

Second, two orientation sessions on internship programs provided relatively rich and contextualized information on the marketing strategies of volunteer internships for study abroad students. Specifically, one orientation was sponsored by a local study abroad agency that catered to South Korean students in Toronto and it was hosted by a local language institute specializing in internship programs. It was conducted in Korean by a Korean agent, and the participants were also Koreans. The other session was an orientation hosted by an internship department at Lingua City. The coordinator, who was a Vancouver-born Canadian, organized and led the session in English with the participation of three German students and one Korean student.

Third, 27 out of 224 field notes and 24 out of 40 interviews with South Korean participants (study abroad agents and students) contained themes and topics pertaining to volunteer internships. Specifically, the field notes and interviews with South Korean students introduced and illuminated the following themes: why they decided to do an internship or not; what they had heard about internships from their friends or study abroad agents; and how they evaluated their internship experiences, if they participated.

The analysis of these three types of data produced the following processes of institutionalized and commercialized volunteer internships in study abroad: (1) intermediaries' marketing of internships as volunteerism (but not paid internships such as co-op programs); (2) intermediaries' systematic strategies for successful volunteer internships; and (3) students' discourse on volunteer internships. All excerpts in this manuscript were originally written in Korean and translated into English by the author unless it is indicated that they were originally in English.

The Intermediaries' Marketing of Internships as Volunteerism

As intermediaries in the private sector, study abroad agencies and language schools were expected to address their clients' needs for English language improvement. During my fieldwork, the study abroad agents that I interviewed had a keen understanding of why South Korean students decided to study English abroad, as well as how the language tourism industry attempted to adapt to their changing needs and motivations for studying English abroad over the past decade. For example, Sangkyung, a South Korean study abroad agent based in Toronto, stated that Korean young adults previously used to travel abroad in the vague name of English learning, "overcoming English fear" or "because their friends and relatives studied English abroad." However, this agent noted that despite the persistence of such motivations, contemporary South Korean students came to Canada "with a clear and specific objective": "they come to develop their *spec*, or so-called future career skills." Sangkyung's statement clearly implies that South Korean students tended to view English language tourism as a component of career development rather than as an opportunity for leisure activities; these students were now more goal-oriented and instrumentally motivated to study English abroad.

To meet Korean students' needs for resume-building activities, study abroad agents and schools offered a variety of language programs and courses that allowed students to navigate and experience different sets of linguistic and cultural skills. Notable among them was a packaged program that combined language skills with non-linguistic skills, such as workplace knowledge (Allan, 2013; Urciuoli, 2008). Internship programs represent this trend. Indeed, all of the study abroad agents whom I spoke with concurred that the internship program was currently the most appealing, as students could not only learn and practice English in the workplace, but also gain work experience in an English-speaking company. It was believed that this would be a valuable form of cultural capital when they returned to South Korea and sought employment in the future. One poster promoted "internship in Canada" as "the key to employment" and "a perfect preparation for employment." Another flyer described an internship as "a program to upgrade *spec* and become more competitive" and "a program for employment." In a workshop about the internship program that I attended, the study abroad agent stated, "[I]nternship is not just for English learning, but also for learning workplace culture" and "internship experience in a local company will help you have a stronger resume and increase your chances of passing the document screening stage." The booklet of volunteer internship programs I obtained from Lingua City (hereafter, the Lingua City internship booklet) outlined three benefits of an internship:

- English language acquisition in the workplace
- A chance to experience Canadian business culture
- A chance to meet and work with Canadians (originally in English)

In the marketing discourse, internships were viewed as opportunities to use the target language in an authentic setting, learn business culture and interpersonal skills in the workplace, and build relationships with English-speaking locals. Thus, internships were constructed as a packaged program for learning English and soft skills for employment.

However, one critical issue that language tourism intermediaries had to address was the legal status of the students. Since the students entered Canada with a study permit, they were required to have a legal status that allowed them to work. Consequently, language tourism intermediaries presented two types of internship programs: paid and volunteer internships. They assisted students seeking paid internships in preparing legal documents such as work permits and social insurance numbers. Additionally, they also connected students with companies that were seeking interns as volunteers. The majority of South Korean students who pursued career-oriented activities were more interested in unpaid and volunteer internships because they offered a greater chance of working in their desired professional positions. In contrast, the majority of paid internship programs were in the customer service sector of the tourism or hospitality industries. South Korean students were informed that although, in theory, students in paid internship programs could work in any business, in practice, the vast majority were only employed as restaurant servers.

While a number of South Korean students showed greater interest in volunteer internships because of no required visa status and wider options, language tourism intermediaries were also required to clarify two aspects to students interested in volunteer internship programs to avoid legal issues. First, they informed the students that they would not be covered by the company's health or property insurance and that they would be required to obtain their own coverage. Thus, volunteer interns were required to sign a waiver of liability at the beginning of their placements, which stated that volunteers "will be liable for any loss or damage caused to any person or property during the volunteer work-placement" (the Lingua City internship booklet, originally in English). This declaration indicated that from a legal standpoint, unpaid interns were not workers but volunteers. Second, they specified that the majority of organizations where volunteer interns would be placed were small to medium-sized businesses. During an orientation session at Lingua City, the coordinator stated that internationally renowned companies had stricter requirements, which meant that internships were typically only available to graduates of Canadian colleges or universities. In another orientation hosted by a South Korean study abroad agency, an agent stated that the positions available to international volunteer interns in a prestigious company like Samsung consisted primarily of menial and repetitive tasks, and therefore were unsuitable for South Korean students seeking professional work experience.

In general, South Korean students did not take the first issue—the waiver of liability—seriously, as they had been required to purchase a study abroad insurance policy before their travel; this had been recommended and mediated by study abroad agencies. However, the second issue relating to the size and reputation of potential companies was deemed crucial as students desired to work for global corporations in the future. They believed that the working experience gained in such companies in Canada would be a great asset in helping them find employment in South Korea upon their return. To address this issue, study abroad intermediaries emphasized that they were mobilizing "experience and expertise to suitably match students to companies in order to enhance English language acquisition in a non-student environment, while exposing students to a work-based understanding of local business practices and culture" (the Lingua City internship booklet, originally in English). That is, even if study abroad agencies or schools were unable to find students volunteer internships in large and well-known companies, they could find students positions that were well-suited to students' interests and future careers.

Furthermore, intermediaries highlighted the advantages of working in a small company such as increased opportunities for teamwork, communication with local colleagues, and English practice in authentic contexts. During an orientation, the volunteer internship coordinator at Lingua City stated:

> Why is this (working in a small and medium-size company) good? You will be working with a small group of people, where everybody will know each other. Everybody will be excited to hear about you, your goals, where you are from, and what you are doing. You can create a more personal relationship with them. You will not only have a mentor, but also more opportunities to talk. At a smaller company, everybody is helping on all aspects of the company. The purpose of internship is for you to become a part of the team. That is a lot easier in smaller organizations. (Originally in English)

In this orientation session, the coordinator emphasized that even volunteer interns would be "a part of the team" in a small company as the colleagues showed personal and professional interests and care. This idea served as an important consideration for potential volunteer interns because international students can struggle to access and participate in local communities as legitimate members (Jang, 2020; Piller, 2016).

The Intermediaries' Institutionalization of Volunteer Internships

Internship programs in the context of language tourism are inherently institutionalized, as they involve the migration and labor policies of multiple institutions. The intermediaries play a crucial role in arranging placements and other administrative tasks to connect international students lacking local knowledge and social networks with local businesses. Similarly, as discussed above, the volunteer internship program is also a result of broader institutionalization. In particular, volunteer internships are a type of niche market created by intermediaries in order to circumvent legal restrictions on international students' employment and to meet students' needs for English learning and global experiences. However, since students as clients of the niche market questioned the marketing discourse of paid internships, language tourism brokers presented volunteer internships as a wise investment—one that could improve students' English language skills and offer them a unique global experience. This process further institutionalizes and even commercializes volunteer internships within the infrastructure of language tourism. That is, study abroad intermediaries introduce and facilitate volunteer internship opportunities for students in a highly systematic manner. I have identified three key strategies developed and utilized by intermediaries: packaging, consulting, and documentation.

The packaging strategy of intermediaries is to create a program that combines business English courses with volunteer internship programs. The packaged program assumes that the failure to achieve expected outcomes in volunteer internship programs is due to a lack of business language, culture knowledge, and soft skills on the part of the students. As seen in the overview of the packaged program described in the program booklet collected at a study abroad fair (see Fig. 7.1), the students who were planning to participate in a volunteer internship were expected to take business English courses as well as general English courses to improve their English proficiency, professional knowledge, and skills prior to working for local businesses:

In this manner, the packaged program presents a study plan for a successful volunteer internship as strategic and methodical. It highlights that students can attain an English proficiency level deemed sufficient for a successful volunteer internship by taking general and business English courses prior to the internship.

The consulting strategy is intended to provide personalized, one-on-one service to students enrolled in volunteer internship programs. The service does not consist solely of a consultation from an agent to identify

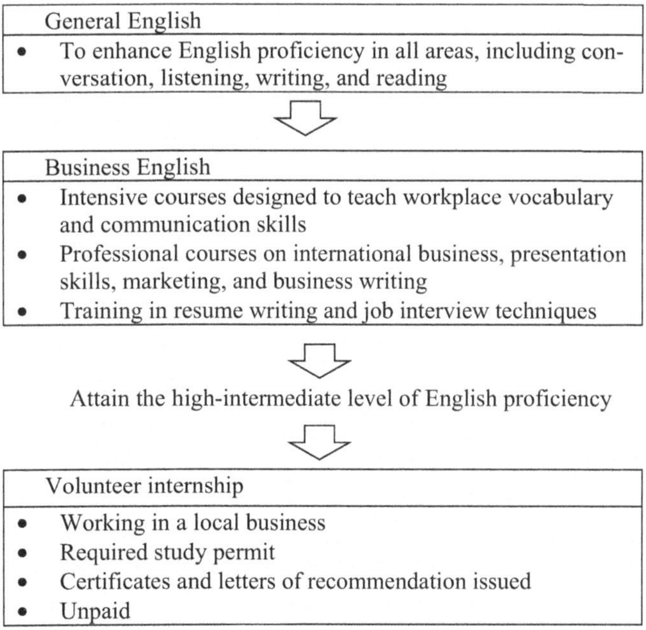

Fig. 7.1 The packaged program for volunteer internships

a student's area of interest and future career for placement. Rather, it also consists of a series of comprehensive, ongoing, and step-by-step consultations until the placement is made and the volunteer internship is completed. For instance, the internship program coordinator at Lingua City arranged the initial one-on-one meeting to discuss a student's placement objective, expectations for the program, areas of interest, and English proficiency. Participating students were then expected to meet with the internship program coordinator to update their resumes while the coordinator searched for a suitable placement for students. When an internship interview with a company was scheduled, a student first conducted a practice interview with the coordinator. Following the interview, the student and the coordinator had a post-interview session in which the student discussed the placement's suitability; the coordinator provided feedback on the student's interview technique (such as language barriers); and they decided whether to accept the placement. They had a pre-placement meeting to discuss basic responsibilities, appropriate attitudes, expectations, and potential concerns. During the volunteer internship, students were expected to report their performance and concerns to the coordinator every week and could contact the coordinator at any time if they encountered an unexpected problem with their employer.

Finally, language schools and agencies required interns to document their experiences in a variety of formats. Lingua City requested that students submit weekly and final reports detailing their activities, feelings, and concerns. The school stated that the information from the volunteer intern's weekly reports would serve as a resource for identifying any problems that a volunteer might encounter on the job, while the final report would allow them to understand the Canadian workplace and track their progress. In addition, the intermediaries informed the students that they could ask their supervisor to write a letter of recommendation, which could reflect how well the student performed in the position. South Korean students were typically interested in receiving a recommendation letter not only because it was not a common type of business document in South Korea, but also because it would serve as an alternative to the certification of completion, which did not detail the duties and responsibilities of the internship position they held.

While a systematic composition of such strategies was intended to encourage internship students to constantly reflect on their learning plans and progress in order to maximize the benefits of volunteer internship programs, it also contributed to promoting the governance of the intermediaries. Indeed, the systematization of volunteer internships necessitated the use of additional human resources by intermediaries. As seen in Lingua City, these programs required internship coordinators who would search for and establish partnerships with businesses, consult with students, and evaluate students' feedback. In addition, language schools that specialized in providing internship services and teaching business English courses—a phenomenon of language industry segmentation—also began to emerge. This commodification of internship programs in the context of language tourism led to the imposition of a higher fee for program registration and internship placement.

The Students' Demystification of Volunteer Internships

During English language study abroad, students frequently discuss what language is "good" and how the "good" language can be acquired, not only because the primary reason for studying a language abroad is to improve it, but also because it is an investment of material and immaterial resources into language learning. While the promise of a volunteer internship for language learning and resume building was made in promotional materials such as flyers, guidebooks, and orientation discourse, South Korean students' perceptions of volunteer internships were mostly formed through conversations with other peers. From my participant observation during the fieldwork, I learned that while students were discussing "good" and "bad" English language programs in casual conversations, the topic of volunteer internships arose frequently. Students talked about not only how to register for the internship program, but also why they wanted or did not want to do internships; what other students experienced during internships; and how they evaluated the efficacy of these programs. These conversations about volunteer internships led to the formation of a dominant discourse regarding the value of volunteer internships among South Korean students participating in this research.

In spite of the intermediaries' marketing discourse and practice, South Korean students' conversations on volunteer internships revealed their shared negative evaluations of internship programs. The reasons included the difficulty in finding a position that corresponded with their majors or interests; the lack of interactions with locals; and the anticipated devaluation of an internship experience in South Korea. For instance, in an interview, Heejun, a male participant, explained why he decided not to participate in a volunteer internship program:

> I was asking myself whether an internship experience would benefit my future career. For example, if I got any position, like working in Tim Hortons (a Canadian coffee and fast-food franchise), that is just a part-time job! I thought that I would not put this experience into my cover letter or resume. It did not help my future career. If I could not find a position related to my university major, I believed that it would be much better to just study at the language school.

Heejun stated that if the position offered by the internship program was a part-time job unrelated to his major, the experience would not be valued for his career development. In addition, his statement that studying English at Lingua City would be more beneficial to him than doing an internship in such a position suggests that the desire for immersion in a local company was trumped by the material calculation of the "return on investment" of English language study. After completing basic English courses at Lingua City, Heejun opted for a Cambridge English Test Preparation course, which was considered the most demanding and academic among Lingua City's English programs, rather than a volunteer internship.

The following remarks by Dongil, another male participant, demonstrate his sarcastic and blunt view of the volunteer internship program. Almost all of the study abroad agencies and schools offered a certificate of completion at the end of an internship as evidence of a student's internship experience. Dongil, however, disregarded the anticipated value of this certification and ultimately did not enroll in an internship program:

> I also wanted to have a certificate of internship completion. Many South Korean students believe that the certification will be valid in South Korea, but I do not agree. I am not certain whether Korean companies will value the certification. From my perspective, it is a kind of garbage. We will receive a certificate of completion when we graduate from Lingua City. I think that would be enough.

Dongil expected that the value of an internship certificate would not be appreciated by South Korean companies. He believed that the certificate value should be determined by the issuer's authority in the market (cf. Jang, 2015).

Another factor that diminished students' enthusiasm for volunteer internships was students' low estimations of the quality of workplace interactions. They frequently heard from their peers that volunteer interns worked alone or that locals were uninterested in them. For instance, Sangwoo, a male participant, stated:

> I have heard that seven out of ten interns work alone. I have also been told that sometimes the intern was the only employee at the company. There were no conversations because most of the work required computer use.

Sangwoo pointed out that establishing relationships with locals is difficult as the work involves few opportunities for communication. Instead of a volunteer internship, he participated in a Korean-Japanese language exchange program because he had a high level of Japanese proficiency. He stated that he had the opportunity to use English significantly more while explaining language differences in the language exchange program.

The following excerpt from the interview with Dongil illustrates that interns' language proficiency and the locals' lack of interest are additional obstacles to communicating with and forming relationships with locals:

> I wanted an internship to meet Canadians, not to do chores. I am aware that experience is important, but two or three months of work experience would not be enough. Above all, I was told that no one spoke with the foreign interns. There are no chances to talk with Canadians. On the top of that, foreign interns often do not understand what Canadians are saying.

Finally, a number of South Korean students complained about the cost of registration for volunteer internship programs. Sangwoo, who declined a volunteer internship, explained:

> The placement fee is excessively expensive. 800 dollars. And no pay! I am aware that experience is more valuable than money. But it is too costly for only two months (of an internship).

Sangwoo was convinced that the fee he should pay to the study abroad agency for a volunteer internship program was too expensive. Furthermore, the fee did not make sense to him because he was not paid for his work.

Due to the financial burden, some students discovered small study abroad agencies that introduced companies without providing customized consultation services and charged a reasonable fee. However, in many cases, they regretted the quality and "fit" of the internship company, or they discovered that the internship did not meet their expectations as discussed earlier.

Such conversations about internships among South Korean students suggest that they demystified the marketing discourse produced by intermediaries and uncovered the illusion of an internship as a valuable experience (cf. Schedel, 2022). In fact, South Korean youth participating in global mobility projects, such as working holidays, confront similar limitations of global working experiences, as these are embedded in racially or ethnically hierarchical labor markets in Western English-speaking countries (Chun & Han, 2015; Yoon, 2014).

Discussion and Implications

This study examined how volunteer internships are institutionalized and commercialized as a study abroad program for employability, and the roles that intermediaries—such as study abroad agencies and language schools—play in the institutionalization of volunteer internships. In the context of language tourism, internships have been perceived as a promising program in the economy of experience; that is, they are believed to provide both immersion and workplace experience for language learning and professional knowledge. Based on this promise of internships, language tourism intermediaries develop volunteer internships to allow students without valid work permits to participate in internship experiences. Volunteer internships are a niche market (Stainton, 2019) within the language tourism industry. As language tourism intermediaries develop and create language education programs and services for the systematization of volunteer internship programs, intermediaries have assumed larger roles, and volunteer internship programs have become more institutionalized and commercialized.

These processes of institutionalization within the Canadian language tourism industry's volunteer internship field are interwoven with various neoliberal discourses and policies and influenced by various stakeholders, as other research on migration agencies has shown (cf. Lorente, 2018; Xiang, 2012; Xiang & Lindquist, 2014). The youth underemployment in neoliberal South Korea and the *spec* discourse that compels South Korean youth to constantly develop skills and qualifications make a volunteer internship experience valuable for enhancing students' employability. In addition, the discourse of immersion for learning a native speaker or authentic type of English (Doerr, 2015, 2018) emphasizes the significance of such global experience. The host country's visa regulations are an institutional element of this educational investment that combines internships and volunteerism. However, the lack of Canadian qualifications and experience and "non-native" English proficiency among international language students make only small or medium-sized businesses available to them. Moreover, as Xiang (2012) notes, intermediaries serve as a central hub in the commercialization of volunteer internships. These intermediaries promote volunteer internships in the language tourism market; address institutional restrictions; and (attempt to) meet students' needs. In particular, they have developed packaged programs, offered personalized consultation, and required students to document their internship experiences so as to maximize the students' returns on this language learning investment (see also Del Percio, 2018; Lorente, 2018).

The increased role of intermediaries in the institutionalization and commercialization of volunteer internships has significant consequences for students' subjectivities and practices in relation to volunteer internships in particular and language tourism in general. Intermediaries' systematic strategies for volunteer internships encourage students to be reflective learners (cf. De Costa et al., 2016; Del Percio, 2018). Additionally, students are urged to create a streamlined plan to attain a sufficient level of English proficiency and participate in a volunteer internship by continuously monitoring whether the plan is followed and goals are achieved through consultations with the internship coordinator and documentations on their practices and emotions. Thus, the intermediaries of language tourism not only mobilize neoliberal discourses of employability (e.g., the *spec* discourse) to make volunteer internships appealing to students, but they also impose neoliberal ethics of self-development and self-direction on students. As Alfonso Del Percio (2018) and Joseph Sung-Yul Park (2010b) point out, this logic of developing and acquiring qualifications and skills frames success and failure as primarily an individual responsibility rather than a social or institutional one.

In spite of the role of language tourism intermediaries in institutionalizing volunteer internships, it does not seem that these internships are fully appealing to international students seeking global experience and native English. Through conversations with peers, such as narrated episodes on the lack of interactions with locals and placement at businesses unmatched to their interests or future career, students demystify the marketing discourse of volunteer internships for English learning and working experience in a Western company. Moreover, the increased commercialization of volunteer internships may make it difficult to gain access to this systematic and personalized service. In reality, a number of students felt a financial burden to enroll in volunteer internship programs due to a higher registration fee, and students who used a brokerage service with a lower quality of volunteer internships often failed to achieve their objectives for immersion in global language and experience. In institutionalized volunteer internships, the quality of the profession and interactions in the workplace, as well as the value of the volunteering experience, is stratified unequally. This stratification occurs despite students being aware of the illusion of volunteer internships and being familiar with the industry's strategies to overcome such limitations.

These effects of institutionalized volunteer internships on students' subjectivities and practices raise an important question regarding who benefits from this infrastructure of volunteer internships in study abroad. This study suggests that those who stand to gain the most from this institutionalized and networked practice of volunteerism are businesses that recruit international students as an unpaid extra labor force, as well as study abroad agencies and language schools that act as brokers for profit (see also Schedel, 2022). In that sense, further investigation is required to analyze (1) how South Korean young adults transform a global volunteer internship into a valuable form of capital in their job seeking processes upon their return to South Korea; and (2) how recruiting companies evaluate students' global experiences. However, given the logic of constant self-development in the neoliberal labor market (Park, 2011), it is questionable whether students' global experiences can be realized as distinctive capital for successful employment.

In a neoliberal society governed by the discourse of self-development and employability, volunteerism is more likely to be regarded as a valuable experience. As volunteering experience is gained by sharing or providing one's own knowledge, skills, and other resources, volunteers can be teachers who aid underdeveloped or deprived communities in the Global South as well as learners who wish to acquire valuable linguistic and cultural capital in an organization in the Global North. However, as volunteers, students would be in a disadvantageous position when searching for volunteer opportunities and engaging in volunteer work as they are more likely to require an intermediary to facilitate the opportunity. Taking the roles and practices of intermediaries as the focus of analysis, this study concludes that intermediaries play a significant role in institutionalizing and commercializing volunteerism. Further research on this type of stakeholders in volunteerism would reveal their roles and strategies, which have been overlooked or concealed thus far in voluntourism and language research.

Acknowledgments An earlier version of this paper was presented in the colloquium, "Between Borders: Language, Brokers and Regimes of Mobility," organized by Beatriz Lorente at the Sociolinguistics Symposium 22. I would like to thank her for inviting me to the panel and introducing me to the concept of migration infrastructure.

References

Allan, K. (2013). Skilling the self: The communicability of immigrants as flexible labour. In A. Duchêne, M. G. Moyer, & C. Roberts (Eds.), *Language, migration and social inequalities: A critical sociolinguistic perspective on institutions and work* (pp. 56–78). Multilingual Matters.

Allan, K. (2019). Volunteering as hope labour: The potential value of unpaid work experience for the un- and under-employed. *Culture, Theory and Critique, 60*(1), 66–83. https://doi.org/10.1080/14735784.2018.1548300

Ambrosini, M. (2017). Why irregular migrants arrive and remain: The role of intermediaries. *Journal of Ethnic and Migration Studies, 43*(11), 1813–1830. https://doi.org/10.1080/1369183X.2016.1260442

Block, D. (2007). *Second language identities*. Continuum.

Brown, P., Hesketh, A., & Wiliams, S. (2003). Employability in a knowledge-driven economy. *Journal of Education and Work, 16*(2), 107–126. https://doi.org/10.1080/1363908032000070648

Brown, P., Lauder, H., & Ashton, D. (2011). *The global auction: The broken promises of education, jobs, and incomes*. Oxford University Press.

Cho, H. (2015). The spec generation who can't say "no": Overeducated and underemployed youth in contemporary South Korea. *positions: asia critique, 23*(3), 437–462.

Chun, J. J., & Han, J. H. J. (2015). Language travels and global aspirations of Korean youth. *positions: asia critique, 23*(3), 565–593. https://doi.org/10.1215/10679847-3125913

De Costa, P., Park, J. S.-Y., & Wee, L. (2016). Language learning as linguistic entrepreneurship: Implications for language education. *The Asia-Pacific Education Researcher, 25*(5–6), 695–702. https://doi.org/10.1007/s40299-016-0302-5

Del Percio, A. (2018). Engineering commodifiable workers: Language, migration and the governmentality of the self. *Language Policy, 17*(2), 239–259. https://doi.org/10.1007/s10993-017-9436-4

Doerr, N. M. (2015). Learner subjects in study abroad: Discourse of immersion, hierarchy of experience and their subversion through situated learning. *Discourse: Studies in the Cultural Politics of Education, 36*(3), 369–382. https://doi.org/10.1080/01596306.2013.871448

Doerr, N. M. (2018). *Transforming study abroad: A handbook*. Berghahn.

Han, H. (2014). "Westerners," "Chinese," and/or "us": Exploring the intersections of language, race, religion, and immigrantization. *Anthropology & Education Quarterly, 45*(1), 54–70. http://www.jstor.org/stable/24029052
Iglesias, M. (2016). The language tourism market system: Conceptualising language tourism. *International Journal of Scientific Management and Tourism, 2*(1), 25–40.
Jakubiak, C. (2016). Ambiguous aims: English-language voluntourism as development. *Journal of Language, Identity, & Education, 15*(4), 245–258. https://doi.org/10.1080/15348458.2016.1195270
Jakubiak, C. (2020). "English is out there—You have to get with the program": Linguistic instrumentalism, global citizenship education, and English-language voluntourism. *Anthropology and Education Quarterly, 51*(2), 212–232. https://doi.org/10.1111/aeq.12332
Jang, I. C. (2015). Language learning as a struggle for distinction in today's corporate recruitment culture: An ethnographic study of English study abroad practices among South Korean undergraduates. *L2 Journal, 7*(3), 57–77. https://doi.org/10.5070/L27323591
Jang, I. C. (2020). The stratification of English speakers in a study-abroad program: An ethnography of South Koreans studying English in multilingual Toronto. *The Canadian Modern Language Review, 76*(2), 155–173. https://doi.org/10.3138/cmlr-2018-0208
Kinginger, C. (2009). *Language learning and study abroad: A critical reading of research*. Palgrave Macmillan. https://doi.org/10.1057/9780230240766
Lorente, B. P. (2018). *Scripts of servitude: Language, labor migration and transnational domestic work*. Multilingual Matters.
Marcus, G. E. (1995). Ethnography in/of the world system: The emergence of multi-sited ethnography. *Annual Review of Anthropology, 24*, 95–117. https://www.jstor.org/stable/2155931
Nelson, L. C. (2006). South Korean consumer nationalism: Women, children, credit, and other perils. In S. M. Garon & P. L. Maclachlan (Eds.), *The ambivalent consumer: Questioning consumption in East Asia and the West* (pp. 188–207). Cornell University Press.
Park, J. S.-Y. (2009). *The local construction of a global language ideologies of English in South Korea*. Mouton De Gruyter.
Park, J. S.-Y. (2010a). Images of "good English" in the Korean conservative press: Three processes of interdiscursivity. *Pragmatics and Society, 1*(2), 189–208. https://doi.org/10.1075/ps.1.2.01par

Park, J. S.-Y. (2010b). Naturalization of competence and the neoliberal subject: Success stories of English language learning in the Korean conservative press. *Journal of Linguistic Anthropology, 20*(1), 22–38. https://doi.org/10.1111/j.1548-1395.2010.01046.x
Park, J. S.-Y. (2011). The promise of English: Linguistic capital and the neoliberal worker in the South Korean job market. *International Journal of Bilingual Education and Bilingualism, 14*(4), 443–455. https://doi.org/10.1080/13670050.2011.573067
Park, S. J., & Abelmann, N. (2004). Class and cosmopolitan striving: Mothers' management of English education in South Korea. *Anthropological Quarterly, 77*(4), 645–672. https://doi.org/10.1353/anq.2004.0063
Piller, I. (2016). *Linguistic diversity and social justice: An introduction to applied sociolinguistics*. Oxford University Press.
Schedel, L. S. (2022). The price of immersion: Language learners as a cheap workforce in Malta's voluntourism industry. *Multilingua, 41*(2), 181–200. https://doi.org/10.1515/multi-2020-0178
Shin, H. (2016). Language 'skills' and the neoliberal English education industry. *Journal of Multilingual and Multicultural Development, 37*(5), 509–522. https://doi.org/10.1080/01434632.2015.1071828
Song, J. (2009). *South Koreans in the debt crisis: The creation of a neoliberal welfare society*. Duke University Press.
Song, J. (2011). *New millennium South Korea: Neoliberal capitalism and transnational movements*. Routledge.
Stainton, H. (2019). *TEFL tourism: Principles, commodification and the sustainability of teaching English as a foreign language*. CABI.
Urciuoli, B. (2008). Skills and selves in the new workplace. *American Ethnologist, 35*(2), 211–228. https://www.jstor.org/stable/27667485
Xiang, B. (2012). Predatory princes and princely peddlers: The state and international labour migration intermediaries in China. *Pacific Affairs, 85*(1), 47–68.
Xiang, B., & Lindquist, J. (2014). Migration infrastructure. *International Migration Review, 48*, S122–S148. https://doi.org/10.1111/imre.12141
Yoon, K. (2014). Transnational youth mobility in the neoliberal economy of experience. *Journal of Youth Studies, 17*(8), 1014–1028. https://doi.org/10.1080/13676261.2013.878791

8

Volunteling the Voluntoured: State-Prompted South Korean English Language and Labor Mobility in Australia

Carolyn Areum Choi

Introduction

> In Ansan, South Korea, there are a lot of migrant workers. But when I came over here, I became those migrant workers. Migrant workers in South Korea do difficult work, but the South Korean migrant workers in Australia are a little different. Australians are thankful to us doing the work that is hard and that they don't want to do. For instance, I was a cleaner at an office. I wiped down the desks in the office and the toilets in the bathrooms, and they always thank me for cleaning their office. When my friends say, "Who is that you are saying hi to?" I say because I clean their office, they are friendly to me. They don't think we are…their servants… In some ways, maybe their approach is "professional?" It makes me have pride in my work.

C. A. Choi (✉)
Princeton University, Princeton, NJ, USA
e-mail: carolynchoi@princeton.edu

L. S. Schedel, C. Jakubiak (eds.), *Voluntourism and Language Learning/Teaching*, Palgrave Advances in Language and Linguistics,
https://doi.org/10.1007/978-3-031-40813-7_8

Twenty-six-year-old Sulli and I were chatting outside of a small park in the middle of Brisbane. She has been living in Australia for the past year as a working holidaymaker with plans to move to Bundaberg, where she will begin agricultural work next month as a condition of extending her visa to an additional year. Despite working various odd jobs as a cleaner, server, and grocery clerk, she distinguishes her work as *different* from the migrant workers undertaking similar jobs in South Korea. She explains, this is because Australians see her as a "professional," not a "servant." To the outside observer, Sulli's comments can be perceived as one explanation for managing the conflicting expectations of her migration experience. While she initially came to Australia with the intention of learning English and gaining work experience, throughout her ten months, she primarily worked under South Korean immigrant employers in the cities and was now going to move to an isolated region of the Australian outback. To some, the dearth of language exposure and practice could be easily dismissed as a "failed example" of the working holiday experience. However, reading between the lines and outside of "failure vs success" frameworks allows us to reinterpret these experiences as ways of making sense and meaning of working holidaymakers' limited exposure to language learning, and more broadly, investing in transformative ways of understanding the contemporary experiences of youth on the move.

Working holidaymakers like Sulli are no exception to the rule but constitute the millions of young people on the move today for language study, education, and cosmopolitan exposure. Youth people represent the world's fastest-growing migrant group with scholars like Shanthi Robertson et al. (2018) pointing out that the current generation of youth are more likely to migrate than any other previous generation of migrants. With mobility now viewed as integral to many young people's transitions to adulthood, scholars have begun to capture the wide array of youth on the move beyond traditional registers of movement. This includes student migrants (Park & Abelmann, 2004; Huang & Yeoh, 2005) but also working holidaymakers (Conradson & Latham, 2005), gappers (Johan, 2009), backpackers (Allon et al., 2008), English teachers (Bernstein, this volume; Collins, 2014), interns (Jang, this volume), and volunteers (Jakubiak, this volume; Schedel, this volume). Such works attend to the multifaceted ways young people navigate the life-course during an era of both increased choice and uncertainty (Robertson et al., 2018; Vrasti, 2013). In recent times, we have also witnessed an empirical shift toward research on Asian youth mobilities (Kawashima, 2010; Tsai & Collins, 2017; Wattanacharoensil & Talawanich, 2018) as the region of Asia has emerged as the largest sender and receiver of educational migrants (Collins, 2013). An emphasis on Asian youth mobilities offers an important window into underscoring intersecting dynamics of youth moves in and from the Global South and non-West, including the roles governments and market competition have played across Asia in facilitating young people's desires for human capital development (Piller & Cho, 2013) as well as more broadly, how questions of race, gender, class, and citizenship stratify migration flows and experiences (Nadal, 2023).

Despite these new directions, the literature continues to echo dominant themes in higher education on value-defined logics of educational or socioeconomic outcomes and returns. This can be partly attributed to the ways shifting neoliberal policies toward education have been often analyzed as a telelogical outcome of capital accumulation and social reproduction. As a result, research on youth and student mobilities may frame migration experiences in a "success vs failure" binary framework that is anchored in instrumentalist and value-defining terms that limit more holistic analyses. These measures tend to focus on whether individuals have "successfully" achieved their intended goals for overseas life. Nancy Abelmann and Jiyeong Kang's (2014) work, for instance, highlights the predominance of "succeeding abroad or failing abroad" narratives in dominant public and media discourses of pre-study abroad experiences in South Korea. The authors point out that such narratives promote the view that *successfully* navigating pre-study abroad not only requires resources and know-how but also the right "character" traits. In challenging these perceptions, they turn towards memoirs of study abroad mothers which complicate simplified depictions of "tiger mothers" and "problematic children." Such works reveal the shortcomings of limited explanatory models while highlighting the nuanced processes and mechanisms of complex experiences that help uncrack the "black box" of youth mobilities.

Works like Abelmann and Kang's (2014) attest to the importance of exploring the myriad ways individuals align or contest these dominant discourses, and furthermore how they make sense and meaning of their everyday experiences. A burgeoning body of work across volunteer tourism, language education, and gender and migration have begun to deconstruct this dominant framing. For instance, Kimie Takahashi's (2013) work examined how the sexual and romantic desires of Japanese women language learners in Australia were more pronounced in their overseas experiences than just their desires for language learning, calling this intersection of desires a "bundle of desires" (Piller & Takahashi, 2006, 2013). Wanda Vrasti (2013)'s work on volunteer tourists in Ghana explored the theme of *persistence* and how—despite disappointments and challenges—some volunteer tourists choose to "stick it out" as a means to recover the narrative of "toughing it out" overseas." Sarah Lipura's (2022) insightful concept of "fringe capital" has been especially key in discerning the different forms of intangible and unexpected forms of value that may exist for mobile subjects, particularly in the context of South Korean students studying abroad in non-traditional and seemingly "less prestigous" educational destinations like the Philippines. These new directions invite us to move beyond *totalizing* accounts of migration and begin to interrogate the "value(s) beyond capitalistic value" (Skeggs, 2014; see Lipura, 2022) of international experiences and education. It has allowed us to focus on the subjective and affective dimensions of international education and travel, emphasizing the experiences of meaning-making, solidarity, and self-development that complicate linear understandings of movement beyond instrumental calculations.

Building on these works, this paper uses in-depth interviews and participant observation with South Korean working holidaymakers in Australia to highlight their subjective processes and experiences of meaning-making and solidarity-building, in ways that move beyond standard instrumental claims and attested desires for language education. In doing so, I illustrate the ways in which South Korean working holidaymakers may encounter transformative experiences of co-ethnic, cross-national solidarity and healing that they argue is missing in their hustle-bustle lives back in South Korea. At the same time, I underscore this discussion by examining the social locations of migrants who *persevere* in their migration journeys, even as they face and encounter disappointments and detours that challenge their initial expectations of life overseas. Interrogating South Korean migrant subjectivities across the ebbs and flows of their migration journey helps us situate their ambitions of "becoming global" in the large political economic context of navigating Australia's racialized labor landscape. Moving beyond "success vs failure" frameworks of international education and travel offers new ways to think about South Korean youths' racialized experiences of struggle and solidarity as crucial to their neoliberal subjectivities as well as their transformative migration journeys.

This chapter begins with a brief overview of the political economic and nationalistic context that structures the overseas experiences of South Korean youth in pursuits of English language skills. Next, I discuss my methods and set up the racialized segregated environment in which South Korean youth enter the Australian working holiday program. I conclude with a discussion of the racialized experiences of the working holiday experience and the ways in which migrant subjectivities towards diverse forms of value and meaning-making contest our existing understandings of language desires and journeys abroad.

South Koreans and Global Travel

As one of the largest senders of working holidaymakers in the world (Choi, 2021), the South Korean case provides a small window for understanding the production of transnational inequalities and new forms of subjectivities among young migrants within the ever-expanding commercial global educational marketplace. The story of South Korean youth mobilities begins at the onset of the new millennium. After decades of insulated economic development under an oppressive military dictatorship and regime, the South Korean government shifted to a global-oriented political economic agenda called *segyehwa*[1] officially, making its move toward economic liberalization in the global marketplace. English education was viewed as a cornerstone of *segyehwa*'s policies (Cho, 2021; Park & Abelmann, 2004), linking English language skills with the

[1] All Korean words in this chapter have been Romanized in the Revised Romanization of Korean version with some exceptions. Note that pseudonyms do not follow Romanization rules.

economic principles of human capital development (Park, 2011; Piller & Cho, 2013). The emphasis on English education paralleled the deregulation of existing educational systems (Jeon, 2009; Park & Abelmann, 2004), ultimately displacing the responsibility of limited state and social services onto surrogate institutions like the family (Jeon, 2009). This shift in education (and English education) from public good to a private household matter ultimately led to the rise of a multi-billion-dollar private English education market at the end of the 1990s (Park & Abelmann, 2004).

The devastation of the 1997–1998 Asian Financial Crisis made way for a new phase of government-sponsored neoliberal (i.e., liberalized market-focused) labor and educational policies (Song, 2009). Under the control of the International Monetary Fund (IMF), South Korea further restructured its economic systems, outsourcing manufacturing production and replacing lifetime employment with part-time, contractual, precarious labor (Song, 2009). It was during this time that scholars of migration began to note the rise of pre-college study abroad to English-speaking countries mostly in the Global North and West but also to the Global South and Non-West (Huang & Yeoh, 2005; Park & Bae, 2009). The scarcity of full-time stable employment had strong implications for everyday South Korean youth, who—despite high levels of high school completion—witnessed record rates of unemployment and irregular employment throughout the mid-2000s and early 2010s (Chun & Han, 2015; Hyundai Research Institute, 2018; Yoon, 2015). Although youth unemployment was widespread, it further entrenched existing class divides (Abelmann et al., 2009), polarizing across the middle classes (Koo, 2007), particularly in the form of regional and educational divisions (Choi, 2022).

The advent of the neo-conservative Lee Myung Bak presidential regime (2007–2013) marked the particular juncture in which youth mobility programs were aggressively promoted by the state. The Lee administration was heavily invested in the production of "the KoGlosian"—a compound word of "Korean," "global citizen" and "Asian"—representing young South Koreans bestowed with the duty of keeping the nation's economy globally competitive (Kim, 2010). The KoGlosian's global expertise was tied to their ability to master English, with one member of

Lee's presidential transition team declaring, "when all [Koreans]...freely speak in English, the GDP of the nation will...rise" (Lee, 2008; see also Chun & Han, 2015). In intensifying English education policies, the Lee administration posed English competencies as a solution to youth unemployment, fueling private spending on both domestic and transnational English language education. Scholar Sujung Kim (2018) remarks that these administrations expanded policies of sending prospective unemployed college students and graduates to diverse forms of low-wage internships, labor programs, and volunteering with minimal pay by glamorizing the potential to become *global leaders*. This is a phenomenon that has been observed in other globalizing societies such as North Ireland and countries in North Africa (see McDaid & Sunyol, this volume; Schedel, this volume).

One of the key projects that expanded was the working holiday program, especially those headed for the English-speaking countries like of Canada, New Zealand, and Australia. The working holiday program is a bilateral migration scheme that allows young people between the ages of 18–30 to enter the country and temporarily live, study, work, and travel. These programs provide individuals with the option of participating in low-wage labor markets (Robertson, 2014), ideally by allowing students to subsidize their overseas language learning experiences while abroad. While Canadian and New Zealand working holiday programs held quota restrictions for South Korean nationals limiting the number of grantees during the time of research, quotas for Australia have increased over time. As Shanthi Robertson (2014) explains, this expansion was directly related to labor deficits in Australia where mostly rural employers pushed for liberalization of the working holiday policy to meet rising labor demands. This is what ultimately allowed Australia's working holiday program to attract the largest numbers of working holidaymakers in the world. And in the eyes of the bilateral governments of South Korea and Australia, this arrangement was a win–win situation: South Korea's ability to export unemployed workers would allow it to maintain its global reputation and gain English language skills at minimal cost while receiving countries like Australia would be able to meet its labor deficit, especially in rural regions of the country, as well as gain tourism dollars.

Methods

This study's findings are based on ethnographic and interview research conducted between 2015 and 2018 in the home country of South Korea and the host countries of the Philippines and Australia. My primary fieldwork was based on participant observation at an English language school in the Philippines and a farming town in Australia.

I gained access to these field sites through a multi-step process. First, I carried out fieldwork at an English language school in the Philippines in 2016. There, I came across the finding that over half of the participants that I interviewed were on a coordinated study-work program known as *yeonge yeonsu* with Australia as the final travel destination before returning home.[2] This is one of the primary reasons I included Australia as an additional field site as a way to understand the complete picture of the language travel experience. The following years, I traveled to Australia to meet with my contacts. I was able to conduct follow up interviews from the original sample from the Philippines by meeting up with them in various cities, including Caboolture, Cairns, Brisbane, and Sydney. Additionally, I interviewed additional participants whom I met in Australia who had come either directly from South Korea or from the Philippines for English language study.

In Australia, I undertook participant observation at a farming town in Bundaberg. There, I stayed with migrants in a boarding house and also worked with them in the fields. In Bundaberg, I was able to undertake additional interviews with South Korean migrant workers, some of whom had come from the Philippines, others who came directly from South Korea. In total, I undertook 41 in-depth, semi-structured interviews and demographic surveys with South Korean working holidaymakers who resided in Australia with interview topics, which ranged from class and family background; migration process and trajectory; educational background; and prior work experiences. The interviews were conducted in Korean and audio-recorded, and I analyzed my data through data analysis software NVivo.

[2] This pattern has been observed by other researchers including Chun and Han (2015) and Jang (2018).

Australia and Racialized Spatial Segregation

There are many different English-speaking working holiday programs in the world. Australia was viewed by participants as the most "accessible destination" with lower barriers of entry to admission and as a place where they could live, work, study, and interact with "native" English speakers. Working holiday visas would typically be approved overnight for South Korean migrants and for a fee of approximately $500 USD. Some admitted that they preferred more "exclusive" working holiday destinations like Canada or New Zealand, which were regarded as desirable due to their particular positionings in the South Korean imaginary but also because of it's more restrictive entry.[3] However, as they explain, settling on Australia as a destination is a happy compromise: having a chance to live and learn English in an English-speaking country despite not having the funds to subsidize a full-length language learning program abroad. On the working holiday visa, working holidaymakers can live and work anywhere in the country for a full year, staying with one employer for up to six months. Their visa allows them to apply to the same jobs as Australian nationals in the formal labor market, offering them as they see it, an invaluable opportunity to "interact" with native English speakers and accumulate cosmopolitan work experiences—all toward their nation-state calls for "human capital development."

Despite access to the formal labor market, working holidaymakers, especially those from non-Western countries, do not often obtain jobs alongside Australians. As Shanthi Robertson (2014) explains, because working holidaymakers are legally designated as "tourists," they are politically hidden as a core part of labor migration policies (see also Ghazarian, 2021). This means their labor is often invisible and unaccounted for in the same way as other groups of labor migrants. In particular, South Korean working holidaymakers concentrate in agricultural, service, and manufacturing industries that rely heavily on temporary migrant labor, which makes up ten percent of the Australian working population (ibid). This includes hotel concierge, hotel cleaning, retail work, food service,

[3] Other Western destinations like Great Britain and the United States were viewed as "study abroad" destinations, which was generally deemed more expensive than the working holiday.

office work, tourism-related work, and regional work including crop-picking and manfacturing work (Kim, 2010; Yoon, 2015). These jobs are largely tied to the low-paid flexible labor market, characterized typically by part-time, temporary, casualized, and contingent labor (Standing, 2011). While many come with high hopes of upward labor mobility and working alongside other Australians, many found that their racialization in the Australian labor market dramatically reduces their employment prospects to manual labor jobs. For instance, Chang, a working holidaymaker from a region in Gyeonggi-do explained:

> *They don't really pick us.* The European friend I went with got picked. That friend is good at English, gentle, and funny. He was Italian, and there was definitely a bias towards European. *They like them more.* If there's an Asian and European, they will for sure use European.

In fact, speaking with participants revealed that in the Australian cities like Sydney and Brisbane, most South Koreans gravitate towards informal off-the-books jobs. Despite their access to the formal labor market, Korean working holidaymakers still find themselves dependent on co-ethnic immigrant employers to access under-the-table cash jobs. These jobs typically include service work in Korean-owned immigrant businesses, such as restaurants and groceries; they are often advertised in popular Korean language community boards that are dedicated to each city, such as *Everyone's Brisbane* and *The Road to Sydney*. It is common for these employers to pay below the legal minimum wage (Yoon, 2015) due to the higher desire for jobs in cities, which migrants see as more ideal locations to live and learn English. Working holidaymakers tend to view jobs in co-ethnic immigrants businesses as an economic lifeline and a means to afford city life despite their low pay. Take the case of Yuji. After realizing that she only had a couple thousand dollars in the bank left to get through the next several months, she explained she had to hustle to find work fast. She said that the easiest jobs to secure were in those in Korean-owned small businesses. For this reason, workers are compelled to take them despite below minimum wages:

> I worked from 10 am to 5 pm every day at the Korean market. After I get off at 5 pm, the city ... isn't far ... by bus but because it's rush hour, there is a lot of traffic. I must be [there] by 6 pm. From 6 pm, it varies, but I worked as a cleaner [then server]. For cleaning, I just have to do ten jobs. No matter how long, when I finish ten jobs, I can leave. It's from 6 pm to ten jobs. But serving at the Korean pub, I have to wait until customers leave. The latest is 11:00 pm to 11:30 pm but as you can imagine, we end late on the weekends, because people stay late. I usually end around 10 pm to 11 pm and go home. I work full-time all day and have no off days. When I go home, I clean my house, and I have to do laundry every day, because Australian migrant workers wear neon or black colors for work. I wear black, so I must wash it every day. After laundry, it's 1 am. Then I sleep, wake up, eat, and go back to work the next day.

Experiences like these demonstrate how migrant workers tend to remain segmented in certain labor markets that are unofficially designated for them. Because of their limited access to more desirable jobs in the formal labor market, many South Korean working holidaymakers, especially those who prolong their stay in the cities, gravitate towards jobs in the informal economy under co-ethnic immigrant employers who may underpay them. For instance, Yuji's cleaning job was for a co-ethnic immigrant employer who, looking to cut corners, did not pay her the legally mandated hourly rate but instead paid her per job. She points out that she was also underpaid below the minimum wage by her employers at the co-ethnic operated grocery store and pub. Such exploitative experiences in the cities can sometimes facilitate migrants' decisions to move to the countryside, especially in agricultural work, where they see more opportunities to make money. This is something that migrants often learn about prior to migration or through informal networks with previous work experience in Australia. These jobs are advertised on Korean-language job listing pages and blogs, promoting a message of higher earning potential due to the system of compensation based on the productivity of work. Together with the possibilities for visa extension, South Korean migrant workers—looking to earn more money and escape experiences of labor exploitation in the cities—will move to "the country," where they encounter a different set of racialized geographical segregation.

The Australian Countryside

> "I earned $500 per day at the cherry farm." "I worked at the farm and after working through the holidays I earned $3000." *You hear about these jackpots.* I was like, "Should I make a lot of money here and do something?" They say you go on an Australian working holiday to make a lot of money. They said if you spend all your money there and come back, you are stupid. That's why I wanted to make a lot of money. (Jin, 25, emphasis added)

Beyond desires to learn English and gain cosmopolitan experiences, a large number of South Korean workers I met were, as Jin remarks, motivated to "make a lot of money" in Australia. These were young people primarily coming from service and tertiary sectors of the South Korea labor market, looking to make earnings that could jumpstart their economic futures back home. Discussion of these jackpots in South Korean circles were linked to certain jobs on the working holiday. Next to better paid jobs, like housekeeping and factory work, farming jobs were viewed as the most easily accessible and lucrative because migrants are paid "per piece" or per bucket, incentivizing higher yields for production. Farming jobs like crop-picking were important in meeting two key needs for South Korean working holidaymakers: (1) it would allow them to fulfill the requirement for the visa extension or "second visa" and (2) it would enable migrants to earn money fast, especially those who do not have the time to wait to find other jobs.

Agricultural labor, however, is channeled through an intricate system organized by South Korean migrant labor brokers. As Jang (this volume) argues, intermediaries play an integral role in the institutionalization of youth mobility regimes and flows. While some workers in my study found labor brokers through educational agencies that pre-arranged work engagements, most workers that I spoke to found jobs through searching Korean language blogs dedicated to certain types of crop-picking that were often region-specific. Labor brokers typically charge "introduction fees" for securing access to jobs that would allow migrant workers to complete the 88 days of work needed for the second visa. These middling industries capitalize on the replenishing flow of South Korean migrant workers who arrive in Australia with minimal networks to find jobs, tethered to conditions of their immigration status. As the go-between for migrant workers, labor brokers are best poised to maximize the spatial segregation of migrant workers in regional sectors, playing a part in the ongoing racialized exploitation of South Korean working holidaymakers in Australia.

Making money in crop-picking is not exactly a straightforward process. Labor brokers can easily manipulate the absence of information and connections of isolated migrant workers in rural areas. Migrant workers often complain about how they would get paid less than their expected pay for the total number of buckets filled, leading to increased suspicion that labor brokers and managers were skimming buckets for profit. This was widely discussed among South Korean migrant worker communities and interwoven in many migrant stories about rural farm life. Take for example 27-year-old Tug, a high school graduate in construction from Jeju City, the capital of the largest island in the South Korean peninsula. He left Jeju Island after completing his compulsory military service. With a disabled single mother and having himself been recently diagnosed with an illness, Tug was eager to leave on Australian working holiday to "stock up on money for his mother" and as his last bucket list item before he "doesn't know what would happen." With a lot of moving parts in Tug's life, he came to Australia in a scramble, landing in the outback within the first few months of his stay to secure the second visa. Despite a strong resolve to make "a lot of money," his plans were cut short when he was conned by a labor broker he found via Instagram:

> I found a Korean brokered farm. It was a grape farm, so I went. But it was already winter so there were no grapes. The pictures on Instagram were deceiving. The Korean broker was a con artist. He took all our money and didn't even give us a second visa. You are supposed to get paid piece price or as much as you pick. One box of grapes is $3. You pick all day. But peak season had passed so there were not a lot of grapes. It was hard to even fill one box. When there are no grapes, it's hard to even make $30 a day. I worked six days a week and earned $180, but my rent was $110 per week. I have $70 remaining. How do I live? But we looked forward to getting our second visa. But the guy was a con artist. We applied for our second visa, but it all got cancelled and that's when we knew we'd been conned.

Tug's experiences illustrate the compounded vulnerabilities migrant workers in farming communities face as they navigate their spatial isolation in rural Australia. South Korean migrant workers are overrepresented in regional work sectors, such as crop-picking, because of the high demand for workers, opportunities for income-earning, and their overall labor market exclusion from other work. Co-ethnic migrant labor brokers, who are often legally, socially, and financially precarious themselves, maximize opportunities for profit among newly arrived South Korean migrant workers as in the case of Tug. As a product of Australian labor policies, the geographic segregation of South Korean migrant workers in rural areas (through their exclusion from other labor markets) reproduces a racialized hierarchy of labor in Australia (see Piller & Lising, 2014). This can ultimately create situations of scarcity and desperation that result in recurring instances of worker abuse—oftentimes at the hands of other co-ethnic migrant workers themselves.

Experiences of "Value Beyond Value"

At the same time, despite instances of structural and co-ethnic exploitation, spatial segregation is not just a one-way street. While life in the farms and factories may be much different than migrants' initial hopes of gaining English language exposure, migrants explain that they gain highly impactful subjective experiences that shape their self-development as individuals. A commonly discussed experience among South Korean working holidaymakers has been developing feelings of solidarity especially with other co-ethnic and racialized working holidaymakers. Upon their arrival in rural Australia, migrants find themselves working alongside other co-ethnic migrant workers in a pod, while interacting with other racialized migrant groups on the fields and in the factories. These experiences of isolation also become sources and moments of co-ethnic and cross-ethnic solidarity formation that help, at times, foster a newfound critical racial consciousness, that moves migrants beyond the language desires that initially motivated them.

Because South Korean labor brokers organize the labor, carpool, and living arrangements of South Korean migrant workers, they are largely segregated together, forming tight-knit communities.[4] Migrant workers who lived in the same boarding house form a pod, where they would go grocery shopping, dine out, and also take trips and excursions together. These tight-knit groups would often convene with the larger Korean migrant worker community (organized under the same labor broker) on a regular basis across work sites. Over time, these communities formed a close bond that many mentioned was an important factor in making it through the hard times in the countryside. For instance, 30-year-old Yohan who came from Northern Seoul and Goyang would often talk about how his worker community was key to navigating life in Bundaberg:

> We became family. The work is so hard. But at least you have your friends you look forward to. They help get you through. We help each other out. When one of the workers needs help lifting, we will go lift it. If one of them doesn't have enough fruit to fill their last bucket, we donate. That's how we get through. That's how we got so close… It's also how some partners got together here.

Yohan's comments illuminate the importance of close bonds in getting through the laborious experiences of the working holiday. Additionally, he alludes to the fact that his close friends on the farm were in the "same situation as he is" in South Korea, often commenting that his friends here are "the country people of Korea in the countryside of Australia." Such findings are echoed in other studies of youth mobilities where friendships are highlighted as a way to demonstrate the instances of meaning-making and value creation forged in often what are difficult and tumultuous journeys abroad (Robertson, 2018; Schedel, this volume).

Scholars have previously discussed how migrants studying and living abroad can often experience enhanced feelings of ethnic nationalism in response to feelings of racial alienation, especially in white settler and Western contexts (Kim, 2013). Ethnic solidarities are furthermore forged through the establishment of co-ethnic migrant spatial formations and

[4] Scholars like Piller and Lising (2014) have noted that migrant workers can also see co-ethnic closeness critically as a constraint to opportunities to learn English. This was also a sentiment mentioned during my interviews; however, for many I spoke with, it did not necessarily discourage them from forming co-ethnic ties.

community. Facilitating the building of community in rural Australia is the establishment of small "Koreatowns" in certain work sites, where South Korean workers can purchase Korean food and products. In the isolated rural farming town of Bundaberg, there was a Korean market that sold *soju*, Korean meats, Korean-branded condiments, and over two dozen types of Korean instant noodles—a staple food for South Korean migrant workers living abroad. Workers would gather together for "family dinners" in their boarding houses, eating Korean food and drinking Korean spirits that many would describe as the emotional lifeline for getting through the days. This is reflected in my field notes:

> In the boarding house where I stayed, my share mates and I would go to the Korean market at least two times a week, especially because we had people who liked to drink in the house. We would make Korean meals almost every night from kimchi soup, army soup, to kimchi pancakes, whatever we could make from the ingredients we could gather at the Korean market. We were luckier than people in other houses, because one of our share mates had purchased a car and didn't have to walk two miles to the Korean market.

Whether in urban or rural areas, South Korean working holidaymakers interact less with what they imagine as "native English speakers" in Australian society. Their social and spatial segregation points to their racialized experiences in Australia, where white settler hegemonic racial hierarchies place them on the margins of the labor landscape. Yet, low-wage labor is not just seen as "failure" of fulfilling language desires by many. The solidarity formation they experience with other racialized working holidaymakers is what also defines their experiential meaning-making as young people pursuing "global citizenship projects." This means that they have managed to carve systems of support from their labor market segmentation and racialized displacement in the most precarious and underpaid jobs.

Their relationships with other migrant workers can also foster bonds and moments of cross-racial solidarity—something that South Koreans coming from a largely homogenous nation may be experiencing for the first time. This was especially true for migrants who were coming from smaller towns outside of Seoul and might not be familiar with the growing migrant populations in the cities. This was the case for Borami in her mid-twenties from Asan. Borami had initially come to Australia with plans to improve her English language skills for the South Korean job market. While she did later come to realize that her interactions with "native English speakers" would be limited in the countryside where she found work, this did not discourage her. Instead, she describes how her interactions with other migrant workers in English still helped her to improve her language skills. But more importantly, she pointed out how practicing English with co-workers was essential in developing friendships forged through tough times on the shopfloor:

> There were people from all over the world. All the managers were, of course, White, but the workers included South Asian, Chinese, Russian, and Ukrainian. There were four other Koreans including myself. I was really close with all of them. We hung out *a lot*. We went fishing, we went out to eat, we went to visit the beaches, mountains, and gardens. On the weekends, we would go out with our co-workers. Some are planning to visit me in Korea.

These friendships were integral in offering South Korean migrant workers a deeper understanding of the larger web of labor exploitation surrounding them. In turn, they offered insights on a form of comparative racial consciousness that helps to provide experiential knowledge on how global inequalities work and the position they are placed in within this system. This was the case with South Koreans who were coming to Australia directly from English language programs in the Philippines, where they were not workers but consumers. This was true for 24-year-old Jeji, who worked as a manager at a fast food burger chain Lotteria in Ansan, a factory hub near the outskirts of Seoul and home to a large migrant worker community. After graduating from a trade college, Jeji continued to work at Lotteria, a job she had held since her final year of high school in order to keep a steady income. Coming from a single mother-headed household, Jeji was able to pay her way through school and was motivated to pay off the rest of her student loans by working in Australia, where she heard that you can receive a big payout. For young people like Jeji, working manual, low-wage jobs in Australia may at times feel like a natural extension of her life in South Korea:

> To be honest, I worked at Lotteria before I came here. If you look at it one way, what we do here is also manual labor and service work. They are both hard jobs… I did come from the Philippines and the position in which I was seeing Filipinos is how Australians are viewing me. But this is a Western country, and it is something that I have to learn. I am actually from Ansan in South Korea, which is a factory town with a lot of migrant workers. The migrant workers I see always have a dark and sad expression. Even on the subways or public transportation, they never sit on the seats. I am afraid that will happen to me.

At first glance, Jeji's racialization signals her social decline in a constructed idea of *social status* tied to her home country. It is a decline magnified by its sharp contrast to her experiences in the Philippines, where she had undertaken an English language study program to prepare her for Australia. In the Philippines, Jeji enjoyed an albeit bounded "middle class" experience as a student-consumer where her Korean *won* stretched farther there. However, upon her move to Australia, like the flip of a switch, her situation and status dramatically changed where she descended into the lower levels of the racial capitalist labor hierarchy. As she explains, in Australia, "the position in which I was seeing the Filipinos is how Australians [were] viewing me." She also alludes to observing other racialized migrant workers in her hometown of Ansan, drawing parallels to how she was racializing others and now how Australians were racializing her. She is one of many migrants who relocate to Australia from the Philippines, where educational brokers organized multi-country education and labor itineraries. Yet she was also able to relate her experiences in Australia with migrant workers in South Korea who she mentions, often have defeated expressions. As a caveat, she explains, "I am afraid that will happen to me." Jeji's comments illuminate the experiences of a budding comparative racial consciousness among South Koreans in Australia. By comparing her own experiences as a precarious worker in South Korea and Australia to the cases of migrant workers in Ansan and local workers in the provincial Philippines, she is able to unlock how the web of racialization in global society operates through a constructed global racial order that dehumanizes people's value and worth based on their racialized and geopolitical position within each society. In other words, South Korean migrant workers may imagine an alternative to these mobility regimes: instead of a linear accumulation of cultural capital, they learn the contours of critique of the same systems that compel their movement.

Paradoxical Inclusions

In between moments of racial consciousness, migrant workers still can struggle with navigating the daily realities of downward racialization. They can find it especially difficult to reconcile the misalignment between their initial desires for language learning and actually realized experiences on the ground. How do non-white migrant workers who relocate to a Western white settler nation make sense of their shifting racialized status? In my research, I found that South Korean migrant workers may find generative alternatives to making sense of their racialized experiences by at times naturalizing the state of the constructed Western racialized order. Late twenties South Korean working holidaymaker Yeerim from Seoul, she explains:

> This is how I think. It's not my country. It's unconscious. I don't even think of it as discrimination. I haven't thought, "Why aren't we treated hospitably?" "Why are we suffering here? Why are we treated so poorly?" Because I just think it's all a given. It's natural (to be treated poorly). Of course, if we interacted with foreigners and got to know their culture and spoke the same language, we would find things in common? Also you tend to be pulled towards people who look like you. I just think that's natural. You are attracted to people like you. It's not that Australians hate and look down on us, it's just because they like people who look like them. They like them [white European migrants] more than us. They do it more for their own [white] people. So I don't think it's an injustice.

Yeerim does not think of her poor treatment in Australia as instances of racism and racial discrimination but rather as *natural* for people to be "pulled towards people who look" like each other. Her rationalization of the xenophobia and racism in Australia seems to reproduce discourses on racialized hierarchies that champion whiteness and naturalize racism. While migrants do engage in cross-racial and co-ethnic solidarities that help create alternative forms of value and forms of collective consciousness, this does not eliminate their subscription to discourses of racialized hierarchies. In fact, some migrants may feel hypersensitive to taking a more empowered stance due to an overwhelming precarity of their conditional status as temporary working holidaymakers. As Yeerim iterates, "This is how I think. It's not my country."

What we begin to notice are the serious limitations imposed on young adult migrants to develop a more liberatory and critical social consciousness as a result of their transnational mobility. While we can identify moments of community formation and consciousness-building, we also find that in many circumstances, due to migrants' precarious racialized and legal positioning as excluded and conditional "non-citizens" in these receiving countries, the potentiality of this often becomes stymied. In turn, it becomes clear that the "global citizenship projects" that many young people are aspiring for becomes an intractable feat, especially for the more precariously positioned in South Korean society. These workers do not fall under the description of what David Jefferess (2008) describes as the "global citizen": [a] "specifically positioned subject that is constituted by the ability to act, and 'make a better world' for… Others" (p. 28). Instead, this group is comprised of subjects who interact head-on with the structures of global inequality and racial capitalism and who try to persist even in the face of many structural and legal barriers and constraints within the migration landscape. In many ways, Yeerim's desires for belonging and acceptance in Australia speak more to the larger troubled implications of state-prompted "global citizenship projects," which have made such journeys compulsory to middle class mobility and transitions to adulthood. Her "paradoxical inclusion" as a "racialized other" is a condition that she accepts for her goals reached of experiencing Australia. Once back in South Korea, it is an experience that will potentially offer (but not guarantee) cultural legitimacy, regardless of her racialized experience.

Conclusion

In highlighting the experiences of South Korean working holidaymakers in Australia, I demonstrate how young people with limited resources challenge existing understandings of the language study abroad experience by engaging in strategic and organically experienced forms of value creation, meaning-making, and solidarity-building. While such experiences have been previously examined in the literature on youth mobilities, they are still often framed within a binary framework of success versus failure. Through ethnographic immersion, this chapter is able to move beyond dominant discourses and frameworks and presents a way for us to better understand the fringe capital (Lipura, 2022) and meaningful experiences that come with language travel and fulfilling language desires. While some might be quick to judge that the case of South Korean migrant workers in Australia may be another case of "unfulfilled overseas youth experiences," I demonstrate instead how turning to the voices and stories of the migrants themselves allows us to observe key subjective processes and experiences that have the potential to go beyond instrumental measures of advancement.

At the same time, this chapter also reveals the critical ways South Korean working holidaymakers experience deep social inequalities and racialization. Their precarious positioning in the Australian citizenship system does not readily enable them to (at least initially) protest this unfair treatment (for exceptions see Robertson, 2013). This can at times also reproduce racialized ideologies as they try to "make sense" of rampant exploitation or racism. Foregrounding their experiences of cross-racial and co-ethnic solidarity has been one way they have been able to persist against the hardships and barriers that they encounter in a racialized and segregated labor market. In many ways, it becomes a strategy to justify the feasibility of this journey and manage the stressful duration of their stay in Australia. This is due to the fact that on the other side, they are hopeful that their experiences in Australia can be recognized in the South Korean labor market as culturally legitimate and legible. Such discourses ultimately show the power of global citizenship discourses and the ways in which they operate within individual desires and motivations. This chapter seeks to illuminate how the racialized experiences of South Korean working holidaymakers are not just a one-way street of racialized downward mobility but can be reiminaged by migrants themselves in ways that recognize the broader value of their experience beyond instrumental calculations.

References

Abelmann, N., & Kang, J. (2014). Memoir/manuals of South Korean pre-college study abroad: Defending mothers and humanizing children. *Global Networks, 14*(1), 1–22. https://doi.org/10.1111/glob.12017

Abelmann, N., Park, S. J., & Kim, H. (2009). College rank and neo-liberal subjectivity in South Korea: The burden of self-development. *Inter-Asia Cultural Studies, 10*(2), 229–247. https://doi.org/10.1080/14649370902823371

Allon, F., Anderson, K., & Bushell, R. (2008). Mutant mobilities: Backpacker tourism in 'global' Sydney. *Mobilities, 3*(1), 73–94. https://doi.org/10.1080/17450100701797323

Cho, J. (2021). English fever and American dreams: The impact of Orientalism on the evolution of English in Korean society. *English Today, 37*(3), 142–147. https://doi.org/10.1017/S026607841900052X

Choi, C. A. (2021). Segmented pathways of educational mobility: English language schools, working holidays, and divergent prospects among South Korea's global youth. *Global Networks, 21*(1), 40–63. https://doi.org/10.1111/glob.12307

Choi, C. A. (2022). Transperipheral educational mobility: Less privileged South Korean young adults pursuing English language study in a Peripheral City in the Philippines. *positions: asia critique, 30*(2), 377–407.

Chun, J. J., & Han, J. H. J. (2015). Language travels and global aspirations of Korean youth. *positions: asia critique, 23*(3), 565–593. https://doi.org/10.1215/10679847-3125913

Collins, F. L. (2013). Regional pathways: Transnational imaginaries, infrastructures and implications of student mobility within Asia. *Asian and Pacific Migration Journal, 22*(4), 475–500.

Collins, F. (2014). Teaching English in South Korea: Mobility norms and higher education outcomes in youth migration. *Youth Migration and Spaces of Education, 12*(1), 40–55. https://doi.org/10.1080/14733285.2013.851064

Conradson, D., & Latham, A. (2005). Ordinary and middling transnationalism. *Journal of Ethnic and Migration Studies, 31*(2), 227–233.

Ghazarian, P. G. (2021). Youth as temporary workers abroad: The experiences of Australia, Canada and New Zealand. In D. Cairns (Ed.), *The Palgrave handbook of youth mobility and educational migration* (pp. 323–332). Palgrave Macmillan. https://doi.org/10.1007/978-3-030-64235-8_29

Huang, S., & Yeoh, B. S. (2005). Transnational families and their children's education: China's 'study mothers' in Singapore. *Global Networks, 5*(4), 379–400. https://doi.org/10.1111/j.1471-0374.2005.00125.x

Hyundai Research Institute. (2018). Weekly economic review. http://www.hri.co.kr/board/reportView.asp?firstDepth=1&secondDepth=1%20&numIdx=30043

Jang, I. C. (2018). Legitimating the Philippines as a language learning space: Transnational Korean youth's experiences and evaluations. *Journal of Sociolinguistics, 22*(2), 216–232. https://doi.org/10.1111/josl.12275

Jeon, M. (2009). Globalization and native English speakers in English programme in Korea (EPIK). *Language, Culture and Curriculum, 22*(3), 231–243. https://doi.org/10.1080/07908310903388933

Johan, N. (2009). Gap year travel: Youth transition or youth transformation? In R. Brooks (Ed.), *Transitions from education to work* (pp. 136–149). Palgrave Macmillan. https://doi.org/10.1057/9780230235403_8

Jefferess, D. (2008). Global citizenship and the cultural politics of benevolence. *Critical Literacy: Theories and Practices, 2*(1), 27–36.

Kawashima, K. (2010). Japanese working holiday makers in Australia and their relationship to the Japanese labour market: Before and after. *Asian Studies Review, 34*(3), 267–286. https://doi.org/10.1080/10357823.2010.508765

Kim, D. G. (2010). *The actual condition of Korean working holiday makers in Australia and directions for the program's development*. The Ministry of Foreign Affairs and Trade.

Kim, S. (2018). Voluntarily exiled? Korean state's cultural politics of young adults' social belonging and Korean students' exile to a US community college. *Higher Education, 76*(2), 353–367. https://www.jstor.org/stable/45116701

Kim, Y. (2013). *Transnational migration, media and identity of Asian women: Diasporic daughters*. Routledge.

Koo, H. (2007). The changing faces of inequality in South Korea in the age of globalization. *Korean Studies, 31*, 1–18. https://www.jstor.org/stable/23720158

Lee, J. (2008). Ideologies of English in the South Korean 'English immersion' debate. In M. T. Prior (Ed.), *Selected proceedings of the 2008 second language research forum* (pp. 246–260). Cascadilla Proceedings Project. http://www.lingref.com/cpp/slrf/2008/paper2397.pdf

Lipura, S. (2022). Fringe capital and the perceived values of Korean educational mobilities (KEM) in Asia: The case of Korean degree-seeking students in The Philippines. *Globalization, Societies and Education.* https://doi.org/10.1080/14767724.2022.2095501

Nadal, P. (2023). How neoliberalism remade the model minority myth. *Representations, 163*(1), 79–99.

Robertson, S. (2013). Campus, city, networks and nation: Student-Migrant activism as socio-spatial experience in Melbourne, Australia. *International Journal of Urban and Regional Research, 37*(3), 972–988.

Robertson, S. (2014). Time and temporary migration: The case of temporary graduate workers and working holiday makers in Australia. *Journal of Ethnic and Migration Studies, 40*(12), 1915–1933. https://doi.org/10.1080/1369183X.2013.876896

Robertson, S. (2018). Friendship networks and encounters in student-migrants' negotiations of translocal subjectivity. *Urban Studies, 55*(3), 538–553.

Robertson, S., Harris, A., & Baldassar, L. (2018). Mobile transitions: A conceptual framework for researching a generation on the move. *Journal of Youth Studies, 21*(2), 203–217. https://doi.org/10.1080/13676261.2017.1362101

Park, J. S.-Y. (2009). The promise of English: Linguistic capital and the neoliberal worker in the South Korean job market. *International Journal of Bilingual Education and Bilingualism, 14*(4), 443–455. https://doi.org/10.1080/13670050.2011.573067

Park, J. S.-Y., & Bae, S. (2009). Language ideologies in educational migration: Korean *jogi yuhak* families in Singapore. *Linguistics and Education, 20*(4), 366–377. https://doi.org/10.1016/j.linged.2009.09.001

Park, S. J., & Abelmann, N. (2004). Class and cosmopolitan striving: Mothers' management of English education in South Korea. *Anthropological Quarterly, 77*(4), 645–672. https://doi.org/10.1353/anq.2004.0063

Piller, I., & Cho, J. (2013). Neoliberalism as language policy. *Language in Society, 42*(1), 23–44. https://www.jstor.org/stable/23483184

Piller, I., & Lising, L. (2014). Language, employment, and settlement: Temporary meat workers in Australia. *Multilingua, 33*(1–2), 35–59. https://doi.org/10.1515/multi-2014-0003

Piller, I., & Takahashi, K. (2006). A passion for English? Desire and the language market. In A. Pavlenko (Ed.), *Bilingual education and bilingualism* (pp. 59–83). Multilingual Matters.

Piller, I., & Takahashi, K. (2013). Language work aboard the low-cost airline. In A. Duchêne, M. Moyer, & C. Roberts (Eds.), *Language, migration and social inequalities: A critical sociolinguistic perspective on institutions and work* (pp. 95–117). Multilingual Matters.

Skeggs, B. (2014). Values beyond value? Is anything beyond the logic of capital? *The British Journal of Sociology, 65*(1), 1–20. https://doi.org/10.1111/1468-4446.12072

Song, J. (2009). *South Koreans in the debt crisis: The creation of a neoliberal welfare society*. Duke University Press. https://doi.org/10.1515/9780822390824

Standing, G. (2011). Workfare and the precariat. *Soundings, 47*(47), 35–43.

Takahashi, K. (2013). *Language learning, gender and desire: Japanese women on the move*. Multilingual Matters.

Tsai, L. L., & Collins, F. L. (2017). Youth and mobility in working holidays: Imagined freedoms and lived constraints in lives of Taiwanese working holidaymakers in New Zealand. *New Zealand Geographer, 73*(2), 129–140. https://doi.org/10.1111/nzg.12149

Vrasti, W. (2013). *Volunteer tourism in the global south. Giving back in neoliberal times*. Routledge. https://doi.org/10.4324/9780203104453

Wattanacharoensil, W., & Talawanich, S. (2018). An insight into the motivations of Thai working and holiday makers (WHMs). In C. Khoo-Lattimore & E. C. L. Yang (Eds.), *Asian youth travellers: Insights and implications* (pp. 15–37). Springer. https://doi.org/10.1007/978-981-10-8539-0_2

Yoon, K. (2015). A national construction of transnational mobility in the 'overseas working holiday phenomenon' in Korea. *Journal of Intercultural Studies, 36*(1), 71–87. https://doi.org/10.1080/07256868.2014.990361

9

"Gaps," Workers with No Schedule: The Making of Casual Workers in Two Northern Irish Boarding Schools

Jessica McDaid and Andrea Sunyol

"The Job of a Boarding Is Not Just a Timetable"

In early September 2021 Jessica arrived at Imperial, one of the two boarding schools for teenagers aged 11–18 in Northern Ireland (NI) where she conducted fieldwork for her PhD. She first went to see the Treasurer, who was the person in charge of staff and organizational matters within the boarding department, and a few minutes into the meeting, the Treasurer picked up their phone and called "Veronica," who they explained would

J. McDaid (✉)
Universitat Autònoma de Barcelona, Barcelona, Spain
e-mail: Jessica.Mcdaid@uab.cat

A. Sunyol
Universitat Autònoma de Barcelona, Barcelona, Spain

University College London, London, UK
e-mail: andrea.sunyol@ucl.ac.uk; andrea.sunyol@uab.cat

L. S. Schedel, C. Jakubiak (eds.), *Voluntourism and Language Learning/Teaching*, Palgrave Advances in Language and Linguistics,
https://doi.org/10.1007/978-3-031-40813-7_9

show her around campus. Shortly after, in walked a tall girl with long black hair, neatly tied back, dressed in a dainty outfit: Veronica ("Vero") Garcia,[1] the Spanish teaching-and-boarding assistant. She and Vero greeted each other warmly and when leaving the Treasurer's office, Jessica realized how Vero carried two phones: one for work and the other as her personal mobile. Vero explained that she had bought a second device in an attempt to separate the two worlds, as some Spanish parents of the study abroad students Jessica was following had her number.

Vero had come to Imperial for a year to improve her English and gain some teaching experience in an English-speaking context. She believed this would make her CV stand out from the rest back in the competitive Madrid labor market and hoped to save some money to eventually move out of her family home through the complimentary food and lodging offered to her as an Imperial intern. She lived on-site in a cottage recently converted from old stables to serve as staff quarters. However, because of Covid, the cottage was repurposed to house 6th form girls (aged 16–18-years-old) to avoid students sharing rooms. Regardless of the pandemic, it was becoming a fairly impossible task for boarding institutions to find staff willing to live on campus. And so, Vero soon came to realize precisely what being a female live-in boarding member of staff entailed: she was a jack-of-all-trades who was always understood to be on-call whether she was on duty or not. When she did have moments of rest, Vero was nevertheless implicitly responsible for a household of overseas teenage residents.

Toward the end of the academic year, when some of the overseas students in the cottage began to return home after exams, Jessica was relocated to stay in the cottage. She was then able to witness the range of tacit labor Vero was performing on a daily basis, and how many tasks seemed to have been assigned simply because of her gender. The gendering process was twofold. For one, the very arrangement of the staff-student living situation was assumed on the basis of child protection protocols. It was considered institutionally inappropriate for the male boarding interns to reside among students (particularly females) and therefore male staff members had their own private living quarters. Traditionally, matrons,

[1] The names of participants and schools used in this chapter are all pseudonyms.

housemasters/mistresses, and wardens are residential boarding staff figures whose rooms are located within the students' domain[2] and who are remunerated in this capacity. However, as the cottage was not staffed as other boarding wings because it was originally not devised for students, Vero was led to take on matron roles without any of the material or status gains that would have come with the role. A second aspect that gendered such forms of labor was that they were often affect-based, of the traditional "mothering" domain. Vero was continuously phoned by the cottage girls and the institution—without concern as to whether she was on duty or not, and even when there were others (particularly males) who were on duty—to do the following: bring food to the sick, wake the girls up and do "lights out" (meaning she herself could never go to sleep before them), make sure they had gone to breakfast, and then off to school, give students access to the cottage during school hours if they had forgotten homework (and make sure they went straight to class again or both the student and she could face being disciplined), provide domestic and healthcare advice, and above all engage in feeling work with the students, such as comfort them in their homesickness and/or worries, as well as settle internal disputes.

Vero was not the only one in this situation. As we will also see through the story of Lotte, a German teaching-and-boarding assistant at the second boarding school site in this study, Chester, all of this "overtime" labor was neither personal to one particular experience, nor was it something restricted to a specific school. Rather, the dynamics of casualized labor that force young people with (low) qualifications and (limited) work experience to take up tasks that stretch beyond their original roles (as teaching-and-boarding assistants) seems to be a recurring phenomenon happening in boarding schools across the UK. What is, however, particular to stories such as Vero and Lotte's is how workloads and tasks are unequally distributed among incoming interning workers—or as they are commonly known within these schools, "Gaps."

With this chapter, we seek to explore "Gaps" as a social category that occupies a liminal space between TEFL tourism (Stainton, 2018), English-language voluntourism (Jakubiak, 2020; Bernstein, this

[2] As they are the designated adults on overnight duty.

volume), working holidaymakers (Choi, this volume), the gap year market (Simpson, 2005), language tourism (Iglesias, 2015), and the experience economy (Pine & Gilmore, 2011), through the stories of two female teaching-and-boarding assistants, whom we call Vero and Lotte. Critical sociolinguistic ethnography (Heller, 2008; Pérez-Milans, 2015; Martín Rojo, 2010) allows us to explore the everyday practices and the narratives of making sense of the institutional positioning of language assistants employed in NI, to expand to contemporary notions of voluntourism, and how volunteer work affects social actors traveling for work within the Global North in relatively privileged and formal settings—as are elite boarding schools. We claim that being institutionally recognized as a "Gap," or a "Gappy," in said contexts exacerbates the precarity especially of young women who embark on work opportunities abroad primarily motivated by language learning, and that such "work-and-learn-a-language" experiences (Allan, 2019; Schedel, 2022) legitimize new forms of youth exploitation in the vast language-and-education tourism market.

Young Traveling Language Assistants as "Gaps"

"*Gap year students in boarding schools? What are they doing here?*" Jessica asked herself when she was introduced to "the Gappys" at lunch on her first day of fieldwork at Chester. Over the course of her year collecting data in NI, Jessica came to know around 13 (a mix of male and female) "Gappys" across Imperial and Chester. This meant that "Gaps" was not simply a salient category, but also an established community. However, there seemed to be an array of stories behind how each "Gappy" arrived in the schools, and what their roles and assigned tasks were.

The phenomenon of "Gaps" in boarding schools needs to be framed within the language learning tourism boom across Europe (Codó et al., 2019) and Asia (Song, 2011), which has brought many affluent middle-class families to invest in trending English immersion stays in the UK and Ireland. This form of youth mobility generates a complex international import-export language tourism market. Hoping to bring natural, effortless English proficiency to those who engage in them, these highly romanticized (Doerr & Taïeb, 2020) educational experiences annually mobilize hundreds of thousands of adolescents and pump millions into the receiving local economies (English UK, 2021; Marketing English in Ireland, 2021), crucially creating a demand for new profiles of educational professionals to cater to the needs of traveling students.

In trying to understand the stories of mobility of these young workers and through speaking to various of them, we questioned whether our participants themselves conceptualize this experience as a gap year per se. It is not a coincidence that they are institutionally recognized as "Gaps," considering the similarities their work abroad shares with contemporary forms of Gap Year experiences (Simpson, 2005), as we will see. However, the ethnographic exploration of Imperial and Chester made it apparent that the category, "Gap," was institutionally produced. There were social processes in place that categorized young incoming workers as "Gaps," a category that sits institutionally as a hybrid role between a language assistant (Codó & McDaid, 2019) and a voluntourist (Jakubiak, 2018), departing from the paths traditionally established by the Gap Year literature.

"Gaps" as they are understood in UK boarding schools are young overseas workers in their late teens or early twenties. Some, such as Lotte, find themselves in a limbo about their "next step" and hope to find answers in the experience of going abroad. Others, like Vero, are educational professionals who quit or renounce jobs for which they are qualified in their home countries to broaden their professional and personal experience. However, they tend to coincide in that they are motivated to go abroad by their desires to improve their level of English and gain a global outlook through work and life experience, as a form of "hope labour" (Allan, 2019; and also see Choi, this volume). Regardless of the particularities as to why they travel, it seems as though working at a UK boarding school holds promising fulfilments for a range of profile types.

The young workers we explore in this chapter are not taking time off in the same way as Gap Year travelers, nor do they fit the profile of the "white saviour" voluntourist (Butcher, 2002; Jakubiak, 2018; Pak & Hiramoto, 2022) who sets on a personal growth experience in

NGOs—typically in the Global South. Moralizing discourses of helping the "other" in need, and humanitarianism while acquiring the moralized outlook of a "good individual" (Coghlan, 2021; Wearing, 2001) cannot explain their mobility either. They travel from and to the Global North to occupy temporary job posts in countries and settings as wealthy—if not wealthier—than their own. They plan to stay abroad for a full school year, which makes their stay longer than typical voluntourist expeditions (Guttentag, 2009), with the possibility of staying on or transitioning into other contract types within education, and their mobilities are never primarily motivated by going on trips, holidays, or touring the host country.

It is also hard to understand their roles as fully voluntary, as they travel with agreed work contracts and ideally get paid to some extent for the work they do, although to unequal degrees even within the same institution and certainly across boarding school sites. As we saw in Vero's vignette, they clearly perform beyond what is considered their teaching-and-boarding assistant role. Yet, "pay" is at times more symbolic than material: food, lodging, esteem, learning a language, facilitating trip opportunities, making friends, or finding a home away from home, can count as retribution.

In short, they go abroad to "save" *themselves*: the work-abroad adventure is imagined as an opportunity to shift and revalorize their personal and professional trajectories by broadening their professional experience and acquiring cultural capital, including English language skills, through a practice that also opens up travel opportunities. Working as a language assistant—or as a sports coach in the case of many male interns—in a boarding school abroad is a way of circumventing and compensating for forms of social reproduction or class mobility to which they have not had access before, which is particularly relevant as they will end up competing in the labor market with those in their generation who *did*, for example, spend a high school year abroad. This easy-to-land work experience is also a way of escaping countries with high unemployment and limited prospects of career development for young people (like Spain, or South Africa) (Graham & Mlatsheni, 2021; Holleran, 2019), insufficient English language skills, or limited university choices in their countries of origin. In fact, an enticing aspect for several of these international "Gaps" is that, should they wish to remain, their experiences can serve as a gateway to relocate to the UK through sports coaching or becoming a permanent language assistant at the school and/or within the area. Their stay in NI, therefore, is imagined to boost their capitals and has the bonus of gaining a cosmopolitan outlook (Weenink, 2008) that upon return to their countries will hopefully be a highly valuable currency in the job market (Choi, this volume).

How "Gaps" access their schools, as we will unpack, often corresponds to how they envision the experience to fit their own goals. Whether they use teaching-and-boarding established programs or go through informal, non-commercialized channels such as brokers is not necessarily dependent on their financial capital (Jang, this volume), but rather speaks to the aspects of their future career and personal development they hope to hone through the experience. Nevertheless, as we will see, this does not always correspond to what the category "Gap" they occupy at the institutions can afford them. In any event, such language learning and work motivated mobilities are being undertaken as acts of self-regulation, both as a form of travel experience as well as education to acquire new capitals and increase one's competitiveness. However, as transpires in our opening vignette, unlike the Gap Year industry (Simpson, 2005), or TEFL tourism (Bernstein, this volume; Stainton, 2018), these forms of language assistantships are, in nature, only partly regulated (see Codó & McDaid, 2019), or not regulated at all, depending on how they have been accessed. What this often results in is a spectrum of tensions between expectations and performances of duties in these spaces across stakeholders.

Living with "Gaps": Ethnographic Explorations of Two Boarding Schools in NI

This chapter draws on Jessica's experiences of living and doing fieldwork at Imperial and Chester, two state-funded (public), local boarding schools in Northern Ireland hosting cohorts of overseas students on their year(s) abroad, over the course of an academic year (July 2021–July 2022).[3] We have used a range of ethnographic data such as observations, field notes, interviews with "Gaps" at various moments during their experience, and conversations with their employers and colleagues, as well as several field documents. Our knowledge of the "Gaps" phenomenon, however, also builds around a body of ethnographic explorations of internationalizing education and Early Study Abroad (ESA) families that we have been undertaking since 2014 in the context of Catalonia.[4] Our aim is not simply to explore how young workers who undertake jobs that are heavily affect and language-work based (Barakos, 2022) are discursively constructed and produced institutionally as "Gaps," but also, crucially, to understand how the social category "Gap" has the effect of framing some of their work as voluntary.

For that, we focus on the trajectories of institutional engagement of two young women who traveled to be teaching-and-boarding assistants in two different boarding schools in NI. Vero was a 26-year-old from Spain, who applied privately to Imperial through a family friend working as a broker between the school and prospective ESA Spanish families. While Vero had never taught Spanish before going to the boarding school, she had a background in kindergarten and primary education, and had experience working in various Spanish schools, including covering as Group Tutor in formal primary education in Spain during the pandemic. In order to improve her working status in Madrid to better contract types, having a certified C1 level of English (CEFR) and teaching experience abroad would be "like a star on [her] curriculum" (interview data).

Lotte was a 19-year-old from Germany who came to Chester through a competitive non-profit foundation. Like Vero, she too was hoping to

[3] This is part of a broader ethnographic project undertaken for Jessica McDaid's doctoral dissertation on young study abroad mobility among Spanish (pre-)adolescents for English immersion in NI boarding schools, with ethical clearance from the Research Ethics Committee at Universitat Autonònoma de Barcelona (CEEAH: 5751). This thesis is funded through an FI grant by the Agència de Gestió d'Ajuts Universitaris i de Recerca (AGAUR) (grant reference: 2023 FI-300224), and is co-funded by the European Social Fund (ESF).

[4] These studies are part of the body of research undertaken through the funded research projects APINGLO (FFI2014-54179-C2-1-P), and ENIFALPO (PID2019-106710GB-I00), financed by MICIN/AEI/10.13039/501100011033, led by Dr. Eva Codó (UAB) and Dr. Ana Maria Relaño-Pastor (UCLM). https://webs.uab.cat/enifalpo/.

improve her English. Having an experience away from home was something Lotte felt was missing for her to do before going to university and the start of adulthood responsibilities. She also felt that living and working abroad without her family would help build her confidence. Lotte envisioned Chester as a one-year experience as she was eager to begin a veterinary degree back in Germany afterwards. Since she had already been assigned a spot on the university program, Lotte explained that, in order to defer, she would be required to go through an agency that could accredit her working and/or volunteering status during her absence.

The two schools where this fieldwork was conducted were, as mentioned, Imperial and Chester (for students aged 11–18). Both institutions cater for day school students within the catchment area (who go home at the end of the school day) and boarding pupils (who reside on campus) coming from across the UK and Southern Ireland (SI), as well as overseas study abroad students.[5]

Imperial, Vero's school, is located in the center of a small historic city. Locally, it is a rather well-known Protestant institution for its high-performing academic and competitive sporting achievements. Despite its long-standing boarding tradition, less than 15% of its 800 students currently board, most of whom are overseas students. Chester, Lotte's school, is situated on the outskirts of a big city and is surrounded by vast grounds. It is well-known across the country for its sporting achievements and the word on the street is that "posh (Protestant) kids" go there. Home to a historical boarding department, around 15% of over 1000 students currently board, and around half of this percentage are international.

The Changing Landscapes of Northern Irish Boarding Schools

NI boarding schools are almost always highly competitive state-funded (public) grammar schools, with very few fully privatized boarding schools remaining there today. In NI, the categorical difference between

[5] These students are majority Asian (mostly from Hong Kong) and European (mostly Spanish, but also a smaller percentage of Italian and German students).

"grammar" schools and "secondary" schools has generically been that the former correlates to traditional methods of learning with a great emphasis on "serious" academic performance and exam results, which are then compared and ranked across other schools of this nature across the UK. The latter, in contrast, are considered to focus more on the individual rhythm and wellbeing of each student, with less pressure placed on academic achievements. In the context of NI, the connotations of academic excellence and the importance placed on competitive sporting (such as rugby and hockey) grants prestige to "grammar schools," regardless of the fact that these are state-funded (public) schools. The competitive selection process via the transfer test students take at the end of their primary (elementary) school education (aged 11–12) is what makes these schools attractive, as getting into a grammar school is widely considered a marker of academic ability rather than financial status, which still remains an important currency for future employers and university applications in NI today.

Most remaining grammar schools in NI simply function as day schools with their enrollments strictly monitored by the Department of Education to fit specific transfer test criteria, but in the case of those with boarding departments, such as Imperial and Chester, they can admit students privately since these are internally regulated. This is due to boarding schools uniquely having a stipulated number of spots that they can reserve for boarding students. Put bluntly, what this means is that students who might not have reached the academic admissions criteria to enter as day pupils can still enroll as boarders assuming boarding fees.

The practice of using the loophole of boarding to "pay yourself into a grammar school" has existed for decades and is no secret among NI families. In fact, a large percentage of the local NI boarders were there precisely through this avenue. This shortcut, however, is also enjoyed by those who live further away in the UK or SI (NI is widely considered the cheapest destination to board and receive a British education) and, in recent years, also by large numbers of overseas students.[6] Overseas students—a term which since Brexit now also includes European

[6] See https://monitor.icef.com/2020/07/british-independent-schools-record-6-growth-in-foreign-students-in-2020/.

students—are highly profitable for boarding departments as they are legally required to pay non-UK resident fees, which are significantly higher than those of UK and SI citizens.

As the profile of the student body changes (currently, based on Jessica's fieldwork and interview data, more than half of the boarding departments in NI are made up of European and Asian students, with few local students joining each year), so do the backgrounds and various educational, linguistic, and affective needs these students have (Williams, 2015; Khan, 2012). The changing face of UK boarding departments now catering to a growing international clientele has also greatly impacted the already wavering profile of staff working within these departments.

The privatized boarding department is run by a small number of roles with strict hierarchies of power. These include a treasurer, who deals with the legalities of the boarding department and staffing; matrons, women who traditionally have nursing training who oversee medical concerns to varying degrees depending on the size and tradition of each institution; heads of boarding, housemasters/mistresses, housemothers, wardens, and interns. However, the live-in condition of boarding jobs makes it hard to reconcile with building and sustaining family life, most particularly for women. Nowadays, with working hour contracts becoming more regulated, as well as more workers appreciating a better work-life balance, there are very few candidates willing to fulfill posts that entail a complete "delivery of the full self" unto the institution long-term, or at all. The result of this can be seen in the visibly changing profile of the boarding intern, where schools are now casting their nets further afield to fit their staffing concerns, packaging these as exciting opportunities for an eager mobile youth that can help fill the multiple gaps UK boarding schools face today.

"Because They're Cheap Labor": An International Workforce

In November 2021, joining the top table for lunch as was customary for staff at Chester, Jessica sat down to an insightful conversation happening among school managers. The discussion revolved around the annual UK boarding school meeting that had just taken place, where topics such as the future of this industry and the various health and safety demands they must comply with were being considered. These types of institutions are unique in the UK education system. For one, all students traditionally pay some initial fees proportionate to that of a publicly funded state school, and teacher wages are paid by the state. However, the boarding department is a self-funded unit of the school that works with boarding fees paid by families opting for onsite food and lodging during their child's education (an annual range of £12.000–£17.000 for UK citizens, and £15.000–£24.000 for overseas students depending on the school). The upkeep of campus grounds and the building infrastructure are almost solely funded through the boarding department and alumni fundraisers.

The lunch conversation turned to the difficulties boarding schools in the UK are facing trying to stay afloat, with many fellow prestigious schools shutting down their boarding departments, remaining open solely as day schools. One of these now defunct boarding departments, the deputy head offered, directing their attention toward Jessica, was "not in good state and they couldn't find staff to keep it up." Jessica took the opportunity to ask how it came about that they hosted so many New Zealand, German, and South African staff, making reference to the vast number of "Gaps" from these places. The immediate answer the Deputy Head jokingly provided was "because they're cheap labor." Their colleague, however, who was in charge of the "Gappys," gave a half smile and added they *are* cheap labor, but that "this was something that had been going on for years. There's always been that pathway between the places and sometimes they announce the job, but sometimes 'Gappys' reach out themselves."

The channel through which "Gaps" arrive at the school (be it informally through personal connections, job applications, or more established work-abroad programs for young people) and their reasons for applying for a position did not seem to change how the institution viewed them. To put it simply, these young "foreigners" with various levels of experience coming for a working-and-learning experience were all institutionally recognized as "Gaps," regardless of the specifics of their working status at the school (e.g., to be a teaching-and-boarding assistant, a classroom assistant, or a sports coach), nor whether they themselves would identify with the "Gap Year" category.

The categorization of foreign newcomers as "Gaps" was a widespread process, to the extent that Jessica herself was also read institutionally as a "Gap"—as someone *international*[7] and young coming through the boarding department, who was always "around," and who became friendly—and therefore seen—with the "Gappys." She saw how, on the one hand, this was helpful for her to blend in and enjoy a community of young international people during her fieldwork, but it also consequently meant that Jessica's figure became infused with certain "Gappy" expectations, and the assignment of various tasks for her to "help out." Being read as a "Gap" has consequences both for the everyday institutional lives of these young workers, and also for their long-term trajectories.

"Contribute Your Skills Wholeheartedly": Regulating the Days and Bodies of Language Assistants

Year-abroad work experiences are regulated through established assistantship programs or private job contracts between the worker and the school which stipulate what the daily routines of the young workers will look like during their stay. These low paid (around £7k to £11k per annum) teaching-and-boarding positions are successful in looking attractive to workers who either have no previous formal experience in the education sector or are in countries where youth employment is highly precarious (CCOO & IEIUA, 2021). As happens in the Gap Year market, positions can be advertised through brochures which construct the young traveling workers as daring "adventurers"—youth who are enticed to set off through suggestive wording: "one year away from home—wouldn't that be a good idea?"; "venture the step of spending a year as teaching-and-boarding ASSISTANT" (capital letters promotional material's own). While tourism and taking time off are generally erased, brochures and job contracts

[7] While Jessica was born in NI, the fact that she was coming from a foreign university and lived in Barcelona was sufficient to earn her this non-localized status.

do romanticize living abroad as a globalizing opportunity (Allan, 2019; Doerr & Taïeb, 2020). They topicalize on developing professionally and personally as the main return from this activity—bearing in mind that some schools do not pay for these positions.

Like language assistantship programs, there is little regulation for teaching-and-boarding assistantships across programs, particularly concerning pay, hours, and contractual obligations (of employee and employer) (Codó & McDaid, 2019). This means that, within a single school site, the "Gap" community can have a range of vastly differing salaries and job specifications, which are usually not openly discussed by the school or "Gaps" themselves as this is often a point of contention. Brexit has further obscured these kinds of contractual relationships, since work visas as Skilled Workers for the UK require a minimum salary of over £20k per annum (https://www.gov.uk/skilled-worker-visa). Because of this, teaching-and-boarding assistants like Lotte could only be recruited under a Charity Worker visa (https://www.gov.uk/temporary-worker-charity-worker-visa), and Chester has been unable to pay her or any of the German language assistants a salary:

Extract 9.1. "You Can't Get Paid"

Jessica: =so you were going to work\ and then found out you were going to volunteer\

Lotte: =but then/ they phoned me and said yeah/ you only\ you can you can't get paid\ it is like it is\ but then I spoke with my parents about it and they say it's a:- it's a: unique chance to go a:nywhere/ make new experience/ and stuff like that\ we have here a nice home\ we get food\ we can live here\ our flat is nice/ we can do whatever we want to/ the school even paid us our flights to our own trip/ for London and for Liverpool\ and yeah\

The shift in circumstances Lotte describes is a reconceptualization of her work-abroad as volunteering or as a gap year. Lotte had never been away from home for more than two weeks growing up, and as she watched her friends go on study-abroad stays, she thought going away as a "member of staff and not as a student" (interview, 09-12-2021) would be more profitable for herself. The visa development left her unsure: "at that moment/ I don't know how I feel or I don't know how I should feel because the whole time it was\ all time they said oh yeah\ we will pay you\ we will pay you/" (interview, 09-12-2021). The change in the type of contractual relationship (paid vs. unpaid labor) forces Lotte and her family to re-evaluate the experience: money is no longer the basis of the exchange, and work is "paid" through the provision of lodging and food. Her year abroad should now be framed as a volunteering experience. The romanticization of being abroad serves as a discursive masking of what could also be simply presented as unpaid work. Similar to narratives of ESA (see Choi, 2021; Lee & Koo, 2006; Lo et al., 2015), we see how parents and organizing institutions mobilize self-knowledge, experience, and maturity as key in the up-skilling process that will allow these young workers to access better work and education opportunities in an unspecified future, often referred to as "later on" (Kang, 2015).

The Gapification of Workers and the Moralities of Compliance

The everyday activities workers can expect to be tasked with are provided as long lists both in the brochures and in contracts, with distinction made between day school and boarding tasks. School engagements often specify that language assistants will have to teach or provide help around their target language lessons (Spanish and German in the case of Vero and Lotte, respectively), both in the day school and as homework support at the boarding facilities in the afternoons. Their contracts leave a space open to aid and provide support to students of different age groups, and across the ability range, such as struggling or gifted students. Whether this is language teaching-based or more generally as a class assistant of other subjects is not directly elicited in documents, and this often results in "Gaps" spending a large amount of their time working with Special Education Needs (SEN) students.

SEN classroom assistantships are, in fact, actual paid positions within the day school and are funded by the Department of Education. However, for students who are still awaiting diagnosis or who do not fall under the

SEN guidelines (but are still considered by the school to need assistance), "Gaps" usually fill in this role as part of their "school tasks." Helping SEN students took up an important part of Lotte and her fellow German teaching-and-boarding assistant, Susi's, daily tasks. In their interviews they described how they often felt uncomfortable in this role as they were not trained in this capacity and were "doing the job that others have studied for" (interview Lotte and Susie, 09-12-2021). They also explained how tiring this work was, and how particularly challenging it was to do in English, which was not their first language.

Contractual documents show a wide range of tasks that are worded in very general terms: for example, that teaching-and-boarding assistants should help by "generally assisting with the implementation of the daily school routine." In fact, vagueness is perhaps the most salient feature across contracts, brochures, and position descriptions. Vero's contract, for instance, illustrates her supervision duties:

Extract 9.2. Supervision Duties on Vero's Contract

- To undertake any tasks relevant to the efficient operation of the Boarding Department or the needs of the pupils, as these may arise.
- To gain the respect and trust of all pupils, and in particular boarding pupils, and support them in their academic and extra-curricular activities.

The tasks themselves and the wording are nebulous and demand maximal flexibility. Importantly, as we will later argue, they are also very hard to quantify in terms of hours but also even assert whether they have been accomplished or not (and by whom and within which standards[8]).

In addition to being dressed professionally ("SMART BUSINESS DRESS" appears in capitals as a requirement on Lotte's contractual

[8] A major source of contention at schools was precisely the fact that the typology of many "Gap" tasks (e.g., study supervision, morning uniform check, or even dealing with an upset student) meant that they were open to interpretation. Unfortunately, when tasks were considered poorly executed by fellow colleagues, the foreignness, and/or youth, and/or gender of the "Gaps" was often problematized. Their (lack of) background training or institutional systemic precarities, for example, were never considered.

documents), assistants are also required to follow an "impeccable business etiquette." This transpires in how "Gaps" should wear their feelings, too, as Vero's contract echoes: "You should be capable of handling the inevitable problems of home-sickness, of disconnection with family, and of settling into a different culture." The line following being: "You will be dealing with children of a broad age range, and in doing so, must be able to display initiative and maturity, and be trusted to act within clear guidelines." The need to specify forms of conduct in brochures constructs the imagined candidate as a young, inexperienced worker who needs to be reminded of how to behave in a professional environment.

As full-time members of staff, they have to abide by the same rules as any other (permanent) worker. As such they are warned in their contract:

Extract 9.3. A Substantial Commitment
The role in the school involves a substantial commitment. Whereas certain hours and duties are indicated, inevitably the job will be one of involvement in the many aspects of boarding school life. However, there is sufficient time for you to pursue studies or other outside interests and obligations.

These warnings preempt any sort of complaint at being assigned tasks or duties which are beyond their language assistantship roles, especially when it comes to their boarding duties. They are told that they are expected to "**contribute their skills wholeheartedly besides the language department and boarding in order to become a vital member of the school community**" (brochure, bold in the original document). Brochures and contracts open the door to lock, in a legally binding document, the affective dispositions and attitudinal response that workers need to display upon being asked to do more than what is included in the already overwhelming task lists. As the opening vignette shows, being a teaching-and-boarding assistant comes with the expectation that one will have "boundless energy, good resilience, enthusiasm, flexibility, and an understanding of boarding school life," by contract (Veronica's). We argue that this profiles a particular type of person, while their insistence on being able to follow guidelines simultaneously de-skills the role of teaching-and-boarding assistants. This, we believe, is the *gapification* of such workers: it constructs their work as volunteer, un-skilled, and not subject to remuneration beyond payment in kind. Instructions like the ones above are also a reflection of the change the boarding school intern profile has undergone, where, contrary to how this role was envisioned and staffed in the past, most of these incoming, overseas workers have not received previous training.

The affective and attitudinal presumptions that are expected from "Gaps" sometimes clashes with the understanding "Gaps" themselves have of their own position and relationship with the institution. This was certainly the case with Vero, who came abroad seeking a working professional relationship with the school, and who most certainly did not conceive of herself as embarking on a volunteering gap year. After feeling underpaid and undervalued for some time, Vero wrote to the Treasurer in December for clarification on her contractual hours as she believed that the number of working hours were not "adding up" to what she was being paid for, and that she often ended up having to work on days that were theoretically stipulated as her time off. The "cold" response created a "change in the atmosphere." Her inquiry was a turning point for the school's relationship toward her, particularly boarding management's.

Based on the institutional understanding of her role at Imperial, Vero was not complying with the most important part of her contractual agreement: she was not giving herself "wholeheartedly" to the school. The moralization of "Gap" jobs goes beyond the written contract; it is at the core of the relationship and the power dynamics existing between language assistants and the school.

Regulating Time off: Shaping Imaginaries of Mobility Through Language Assistantship Contracts

The moralized logics that govern assistant-school relationships also dictate "Gaps'" personal mobilities according to boarding department expectations of their roles. Teaching-and-boarding assistant contracts walk the tightrope of trying to remain alluring to an international body of workers, while at the same time making sure the school's needs are covered:

Extract 9.4. "You May Also Take Time to Travel Afield"
Every effort is made to provide opportunities for you to travel and see as much of the country as possible. This is facilitated by participating in class field trips and excursions. You may also take time to travel further afield during holiday periods when such activities are normally undertaken. (Veronica's employment contract)

The school presents assistance in trips and excursions as part of the duties that workers can capitalize on by presenting it as an "opportunity" to travel. Unusually, they are also given advice on how to spend their free time, to the extent that their contractual agreements also state guidelines on the regulation of smoking, drinking, use of drugs, and arrival times in the evenings, further creating an imaginary of language assistantships as a work-and-travel experience. This clearly constructs them as working tourists, eager to explore the country and potentially use their post as a base to then "travel further afield." The profile is also of someone who needs to be told when "such activities are normally undertaken," and who, consequently, needs to be given permission to leave campus:

Extract 9.5. "Tours Can Be Arranged Where Necessary"
You are expected to reside on the premises during term. Special periods away, such as tours etc., can be arranged where necessary through the Headmaster or the Treasurer. (Veronica's employment contract)

Tourism, in these documents is not only shown as a possibility, but as something that the position accounts for, and for which "Every effort is made." However, as we can read between the lines of these extracts, it is up to the school's discretion as to where, when, and how this tourism can be enjoyed. It is worthy of note that, in practice, the class field trips and excursions mentioned above meant that the assistants' participation in these would be in the capacity as a member of staff on duty. It was often also the case that, as we saw with Vero's correspondence with the Treasurer and in the opening vignette, even if "Gaps" do not have an established duty set on a particular weekend or bank holiday, they are nevertheless expected to be on call, meaning that, should they wish to leave school premises for that period, they would have to negotiate permission well in advance. In most cases, unfortunately, the working week was generally so intense that "Gaps" either did not have time off or were too tired to make plans, travel, or even find opportunities to work on their English.

The (im)mobility of "Gaps" was not simply physical. In the specific case of Vero, it was career-affecting too. She had been told by the school that they would issue a working visa for her, which she never received. As she found out halfway through her stay from a fellow Imperial Gap who left the job early, it was never actually the school's plan to even issue one for her.[9] The consequences of not having a visa during and after her stay were numerous and put Vero (and others) in a vulnerable situation. She could not open a standard bank account or have a social security number: ultimately her working status in the country was completely illegal.[10] Whenever she had to go back home, she opted to fly through Dublin Airport to avoid UK border control, especially as a fellow Imperial "Gap" did in fact get deported at the airport due to their lack of a working visa from the school.[11] Aside from the anxieties associated with working illegally in a foreign country, for Vero, the school's priority to save money meant that this year of employment would not count toward her contribution to social security. This made her ineligible for unemployment benefits upon return and did not add up to her retirement pension.

[9] This, we can only assume, is potentially due to the fact that, with her salary, Vero would never have met the legal requirements to obtain one in the first place, and perhaps the school was afraid their offer on a charity visa (like Lotte's at Chester) would not look appealing to applicants.

[10] This also meant that they could not register with a doctor there, which generated a lot of anxiety among them about getting sick.

[11] It was often the case that when "Gaps" expressed concern about their visa-less status coming into the country after holidays, Imperial would tell them to "just say you are visiting family."

Gap Years, Voluntourists, or Language Workers and Learners?

The stories of Lotte and Vero show how contemporary forms of work mobility which were originally not devised as unpaid are being framed as volunteer work. As former intern job positions in NI boarding schools are becoming increasingly occupied by young mobile workers, the boundaries between language assistantships, forms of TEFL voluntourism, study abroad and Gap Year mobilities are converging (Jakubiak & Schedel, this volume) and shaping the institutional expectations that are cast over the young workers, with the effect of casualizing their contractual relationships with schools. We have sought to show how this form of language worker—originally employed to support foreign language teachers—is subject to staffing pressures that cause them to become institutionally recognized as Gap Year (voluntary) travelers.

We claim that the category "Gap," and even more so "Gappy," as they are often called familiarly, is not only infantilizing, but also brings them further precarity. Implicit in the ways in which they are patronized and regimented is an understanding that they are not responsible professionals, but rather inexperienced 20-something-year-olds who are more interested in traveling and "having a good time" than in taking their jobs seriously. Such infantilization provides a discursive framework to justify the violation of basic workers' rights: it is assumed that they do not need real money, unemployment benefits, or contributions to their retirement savings. We have seen how brochures and contracts aim to regulate not only the jobs, but also the bodies (Wu & Del Percio, 2019) of the language assistants, as they provide a framework of hyper-regimented affects (Hochschild, 1979) to establish what is considered appropriate and inappropriate behavior, what is morally acceptable and accepted, and how they need to perform to be always ready and happily disposed toward any needs both in school and boarding. In an institution that educates and hosts teenagers, invoking this as the imaginary of attitudes and social activities that language assistants will partake in situates them as an extension of their teenage student body.

We also believe that the precarious terms of employment of "Gaps"—both in economic return and working conditions—force them to find ways to discursively justify why they put themselves in such exploitative work relations. A way of doing this is through positivizing the experience by emphasizing what they get out of it, as Kori Allan (2019) also finds (see also Choi, this volume). They draw from discourses which are widely circulated by ESA participants and their families, as well as voluntourism travelers, and their own employers: they present themselves as both workers and learners, and put forth what they get in exchange (i.e., "growing" from the experience of being abroad, better English, enriched stocks of cultural capital and a spiced up CV) to avoid the category of precarious worker, which would misalign them from neoliberal narratives of self-accomplishment and success. As previously argued (McDaid & Sunyol, 2022), it is precisely the process of reconceptualizing the ups and downs of one's mobility into packaged narratives that allows them to exchange these capitals.

Assistants, who are often qualified educational workers, are also de-skilled and subjected to working conditions which do not recognize their forms of expertise. As we have seen, the phrasing in contracts, brochures, and the ways in which these workers are asked to participate in institutional life opens the window to vague interpretations that make specific tasks or the disposition to work under certain conditions seem as something that they (should) do voluntarily and with "wholehearted" gratitude for the experience they are being given, as Larissa Schedel (this volume) also observes. The direct consequence of these ideological moralistic discourses is that the institution exempts itself from any responsibility over the experience of these incoming workers, such as (extra) payment or having to abide by work regulations. This is because being hired as "inexpert," and the construction of the school and boarding facilities as the space that "provides" for the expertise can justify low salaries (similar to findings by Codó & McDaid, 2019), particularly considering the kinds of responsibilities and tasks they are asked to undertake, and the number of hours they are on duty. This, in turn, has the knock-on effect of being put in a position where they are not eligible for a salary—and turning the teaching-and-boarding assistantship position they signed up for into a volunteering experience.

Linguistic repertoires and gender further stream teaching-and-boarding assistants into specific institutional roles that make them undertake a greater amount of unregulated, affect-based tasks, and work for longer hours. Language assistantship positions to foreign languages taught at the schools are overwhelmingly filled by women, while men—who mostly came from South Africa and New Zealand—were preferably employed in sports. Moreover, as was evident in the case of Vero's living and working conditions in the cottage, simply being a woman in spaces where child protection is of utmost importance, institutional protocols assign differing working capacities for males and females. Traditional stereotyped gendered tasks are not only reproduced but are in fact institutionally produced and catered for. However, these "homemaker," pastoral, and affective-based tasks are neither explicitly remunerated nor acknowledged, precisely due to the gapification of their positions and the gendered moralistic expectations of what "giving yourself wholeheartedly" implies. Therefore, gender seems to intersect both task allocation and their corresponding moral expectations, with immobilizing consequences for the personal and professional trajectories of female "Gaps" during and beyond their employment.

We believe that the stories of Vero and Lotte add to the growing body of ethnographic work on emerging forms of language, work, and tourism mobilities of which this volume boasts. We sought to shed light on the expectations concerning the off-duty time of our sojourners, and similar to findings by Carolyn Choi (this volume), living in close quarters with our informants brought unique insight on the texture of their daily experience within these roles, bringing a gender perspective on institutionally produced inequality as it unfolds in this field. Our study opens questions regarding the discursive construction of the "other" in processes of voluntarization of labor, which we have not been able to address due to spatial limitations but would further our understanding of contemporary forms of voluntourism.

References

Allan, K. (2019). Volunteering as hope labour: The potential value of unpaid work experience for the un- and under-employed. *Culture, Theory and Critique, 60*(1), 66–83. https://doi.org/10.1080/14735784.2018.1548300

Barakos, E. (2022). Language work and affect in adult language education. *Journal of Sociolinguistics, 26*(1), 26–44. https://doi.org/10.1111/josl.12517

Butcher, J. (2002). *The moralisation of tourism: Sun, sand…and saving the world?* Routledge.

CCOO, Gabinete Económico & Instituto de Economía Internacional de la Universidad de Alicante. (2021). *La precariedad laboral en España, una doble perspectiva*. Confederación Sindical de CCOO.

Choi, J. L. (2021). The student as an enterprising self: Neoliberalism, English and early study abroad. *British Journal of Sociology of Education, 42*(3), 374–387. https://doi.org/10.1080/01425692.2021.1888697

Codó, E., & McDaid, J. (2019). English language assistants in the 21st century: Nation-state soft power in the experience economy. *Language, Culture and Society, 1*(2), 219–243. https://doi.org/10.1075/lcs.00017.cod

Codó, E., Sunyol, A., & McDaid, J. (2019, June). Immersion as branding: A preliminary analysis of immersion discourse in promotional materials in education. In K. In Petit (Ed.), *Immersion education: A social approach*. Paper presented at IV EDISO conference. Santiago de Compostela.

Coghlan, A. (2021). Volunteer tourism: It's no use going back to yesterday, because I was a different person then. In R. Sharpley (Ed.), *Routledge handbook of the tourist experience* (pp. 262–273). Routledge. https://doi.org/10.4324/9781003219866-23

Doerr, N. M., & Taïeb, H. D. (2020). Affect and romance in study and volunteering abroad: Introducing our project. In N. M. Doerr & H. D. Taïeb (Eds.), *The romance of crossing borders. Studying and volunteering abroad* (pp. 19–50). Berghahn Books.

English UK. (2021). What is the value of ELT to the UK economy? Retrieved April 16, 2023 from https://www.englishuk.com/facts-figures#value

Graham, L., & Mlatsheni, C. (2021). South Africa's high youth unemployment: Structural features and current responses. In A. De Lannoy, M. Langa, & H. Brooks (Eds.), *Youth in South Africa: (in)visibility and national development* (pp. 131–156). The Mapungubwe Institute for Strategic Reflection (MISTRA). https://doi.org/10.2307/j.ctv2z6qdwp.13

Guttentag, D. A. (2009). The possible negative impacts of volunteer tourism. *International Journal of Tourism Research, 11*(6), 537–551. https://doi.org/10.1002/jtr.727

Heller, M. (2008). Doing ethnography. In L. Wei & M. Moyer (Eds.), *Research methods in bilingualism and multilingualism* (pp. 249–262). Blackwell. https://doi.org/10.1002/9781444301120.ch14

Hochschild, A. R. (1979). *The managed heart.* University of California Press.

Holleran, M. (2019). The 'lost generation' of the 2008 crisis: Generational memory and conflict in Spain. *Journal of Sociology, 55*(3), 463–477. https://doi.org/10.1177/1440783318817907

Iglesias, M. (2015). Language travel demand: New insights into language tourists' perceptions. *Procedia—Social and Behavioral Sciences, 199*, 149–156. https://doi.org/10.1016/j.sbspro.2015.07.499

Jakubiak, C. (2018). Study abroad and volunteerism: Organizational and administrative issues. In J. I. Liontas (Ed.), *The TESOL Encyclopedia of English language teaching* (pp. 1–6). Wiley-Blackwell. https://doi.org/10.1002/9781118784235.eelt0138

Jakubiak, C. (2020). "English is out there—You have to get with the program": Linguistic instrumentalism, global citizenship education, and English-language voluntourism. *Anthropology & Education Quarterly, 51*(2), 212–232. https://doi.org/10.1111/aeq.12332

Kang, Y. (2015). Going global in comfort: South Korean education exodus in Singapore. In A. Lo, N. Abelmann, S. A. Kwon, & S. Okazaki (Eds.), *South Korea's education exodus: The life and times of early study abroad* (pp. 125–146). University of Washington Press. http://www.jstor.org/stable/j.ctvct024z.11

Khan, S. R. (2012). *Privilege: The making of an adolescent elite at St. Paul's School.* Princeton University Press.

Lee, Y.-J., & Koo, H. (2006). "Wild geese fathers" and a globalised family strategy for education in Korea. *International Development Planning Review, 28*(4), 533–553. https://doi.org/10.3828/idpr.28.4.6

Lo, A., Abelmann, N., Kwon, S., & Okazaki, S. (Eds.). (2015). *South Korea's education exodus: The life and times of early study abroad.* University of Washington Press. http://www.jstor.org/stable/j.ctvct024z

Marketing English in Ireland. (2021). ELT industry of Ireland forecasts 80% decrease in revenue for 2020. https://mei.ie/elt-industry-of-ireland-forecasts-80-decrease-in-revenue-for-2020/

Martín Rojo, L. (2010). *Constructing inequality in multilingual classrooms.* De Gruyter Mouton. https://doi.org/10.1515/9783110226645

McDaid, J., & Sunyol, A. (2022, July). *Unlocking children's full global potential in Ireland: Destabilising the local-global binary in early study abroad*. Paper presented at the 24th sociolinguistic symposium. Universiteit Gent.

Pak, V., & Hiramoto, M. (2022). "Itching to make an impact": Constructing the mobile Singaporean voluntourist in Instagram travel narratives. *Social Semiotics, 32*(3), 332–354. https://doi.org/10.1080/10350330.2020.1766263

Pérez-Milans, M. (2015). Language and identity in linguistic ethnography. *Working Papers in Urban Language and Literacies, 158*, 1–26.

Pine, B. J., II, & Gilmore, J. H. (2011). *The experience economy* (2nd ed.). Business Review Press.

Schedel, L. S. (2022). The price of immersion: Language learners as a cheap workforce in Malta's voluntourism industry. *Multilingua, 41*(2), 181–200. https://doi.org/10.1515/multi-2020-0178

Simpson, K. (2005). Dropping out or signing up? The professionalisation of youth travel. *Antipode, 37*(3), 447–469. https://doi.org/10.1111/j.0066-4812.2005.00506.x

Song, J. (2011). Globalization, children's study abroad, and transnationalism as an emerging context for language learning: A new task for language teacher education. *TESOL Quarterly, 45*(4), 749–758. https://www.jstor.org/stable/41307664

Stainton, H. (2018). TEFL tourism: The tourist who teaches. *Tourism Geographies, 20*(1), 127–143. https://doi.org/10.1080/14616688.2017.1298151

Wearing, S. (2001). *Volunteer tourism: Experiences that make a difference*. CABI.

Weenink, D. (2008). Cosmopolitanism as a form of capital: Parents preparing their children for a globalizing world. *Sociology, 42*(6), 1089–1106. https://doi.org/10.1177/0038038508096935

Williams, R. (2015). Coming to terms with our "Asian invasion": A practitioner's perspective on the Korean education exodus in a Christian school setting. In A. Lo, N. Abelmann, S. A. Kwon, & S. Okazaki (Eds.), *South Korea's education exodus: The life and times of early study abroad* (pp. 125–146). University of Washington Press. http://www.jstor.org/stable/j.ctvct024z.11

Wu, M., & Del Percio, A. (2019). Serving people: Consumer economies, classed bodies and the disciplining of female labor. *Language, Culture and Society, 1*(2), 194–218. https://doi.org/10.1075/lcs.00016.wu

10

Afterword: The Wages of Global Experience, Post Unit Thinking, and Post Native Speaker Ideologies in Volunteer Tourism

Neriko Musha Doerr

The Wages of Global Experience

"The Wages of Global Experience," as I call it, are what the volunteers discussed in this volume are getting. In a very different context, W.E.B. Du Bois once argued that White workers' low wages were compensated by a "public and psychological wage" (2021/1935, p. 841)—the pleasure of Whiteness, as their politicians and police came from their people, and institutions were built for them. The White working class fashioned their identities as "not slaves" and "not Blacks," forgetting their shared interests with the Black poor. Drawing on Du Bois' work, David Roediger (1991) connected the formation of the White working class and the systematic development of their sense of Whiteness by highlighting the agency of the working-class people rather than them being manipulated into

N. M. Doerr (✉)
Ramapo College of New Jersey, Mahwah, NJ, USA
e-mail: ndoerr@ramapo.edu

L. S. Schedel, C. Jakubiak (eds.), *Voluntourism and Language Learning/Teaching*, Palgrave Advances in Language and Linguistics,
https://doi.org/10.1007/978-3-031-40813-7_10

becoming racists. Although the contexts as well as the effects of such wages and their resultant violence are very different, this economic metaphor is useful in thinking about the practices and perceptions around voluntourism/volunteering discussed in this volume. To borrow this theorization here is not to trivialize the effects of racism derived from the Wages of Whiteness but to seek to honor its theoretical approach.

Volunteering is usually seen as an act done without expecting anything in return. However, researchers suggest returns for doing volunteering, from strengthening one's resume (Gray & Campbell, 2007; Heath, 2007; Munt, 1994) and personal growth (Gray & Campbell, 2007; Munt, 1994) to merely feeling good about oneself (Jakubiak, 2012) as in "false generosity" (Freire, 1997/1970), as will be detailed. To this list of what volunteers receive in return, I add "global experience," as in the cases of the volunteers discussed in this volume.

The explicit monetary aspect—wages—is important because the cases in some chapters in this volume involve volunteers who are not only uncompensated but even pay money to do the work, which creates a strange situation of "purchasing" the volunteer experience along with the things that come with it (it is strange yet we are so used to it). Here, what they are paying for is what they get in return, which includes not only feeling that they are "doing good" but also global experience, as discussed in the chapters by Cori Jakubiak and Joshua Bernstein. They examined voluntourists who paid money to volunteer to teach English, in the meantime gaining global experience and "adventure," which was how the lessons for obtaining the Teaching English as a Foreign Language (TEFL) certificate before volunteering were advertised (Bernstein). Aviva Sinervo's case especially highlights the voluntourists purchasing the sense of "doing good," as they felt bad for not doing enough volunteer work and thus found their own way to do more work. Although this is probably a matter of them being conscientious and trying to do their best to help others, the fact that they have paid to do so can lead to an interpretation of their efforts as "getting their money's worth."

Cases in other chapters show volunteer works that are not humanitarian but merely unpaid work, often in exploitative labor conditions. These volunteers put up with such exploitation because they get something in return. Larissa Schedel's chapter showed a man from Maghreb (Northwest Africa) who carried out volunteer work for its giving him the possibility to stay in Italy and then Spain without legal documentation, so that he could get out of the conservative lifestyle of his homeland, as well as for becoming a neoliberal subject branded as having global experience. In Chull Jang's chapter discussed study abroad students from South Korea staying in Toronto, Canada, who preferred unpaid internships/volunteer work that matched their career goals over paid work that tended to be menial and offered few linguistic interactions in English with others. What these study abroad students received in return for their unpaid internships/volunteer work was "global work experience" that adds an edge to their resume when applying for jobs in South Korea. Carolyn Choi's chapter illustrated how South Korean working holiday makers put up with discrimination and racialized workplaces in Australia because they were gaining a "global experience" valued back home in South Korea. Though the cases Choi discussed are not quite volunteering because they are paid jobs, they show how these workers put up with unfavorable work conditions—discrimination and racialization—because of the global experience they got in return. Jessica McDaid and Andrea Sunyol's chapter showed that boarding schools in Northern Ireland exploited language assistants from abroad by framing their work as "global experience" for which they should be grateful. Here, the global experience compensated them for their free labor (i.e., work during their off-duty time), acting as a metaphorical wage.

This type of unpaid work that is "paid otherwise" does exist in various contexts outside of volunteering. Free music or acting performances "for exposure" or unpaid internship work "for work experience" are such examples. The employers are actually getting merely free labor in most cases. Any performance, paid or not, is exposure, and any work can be a good (learning) experience, so not paying for such work is an unbalanced exchange, thus unethical. Paying with "global experience" in the volunteer work discussed in this volume can be flat out considered unethical and exploitation, reminiscent of the classic Marxist notion of the exploitation of the proletariat who put up with low wages because they can survive with them, regardless of whether they deserve more pay or not. The cases of volunteers I discuss here are one example of this.

This structure of volunteers being paid the Wages of Global Experience is shaped by three discourses—globalization, immersion, and volunteering as receiving—which I discuss in the next three sections to show further understanding into these discourses in light of the framework of "Wages of Global Experience."

The Discourse of Globalization

The discourse of globalization valorizes the "global experience" in the way that it can serve as a wage. Analyzing the discourse of globalization that became prevalent in the post-Cold War 1990s as an ideology of globalism, Anna Tsing (2000) suggests its three features: (1) futurism that names a specific era (i.e., 1990s on as that of globalization) and predicts its progress into the future, rendering places that are not "globalized" as backward; (2) the conflation of various situations (from corporate globalization to populist or cultural globalization) as all the same globalization process; and (3) the emphasis on breaking down barriers without recognizing conditions that encourage or discourage global flows and the valorization of such breaking down without acknowledging its negative aspects.

The third feature can be understood further with the notion of the Regimes of Mobility suggested by Nina Glick Schiller and Noel Salazar (2013), who point out the differentiated celebration of global flows. For example, the mobility of the (upper) middle class, such as that of business CEOs and study abroad students, is celebrated and encouraged, whereas the mobilities of lower classes such as immigrants and refugees are seen as problems and are thus hindered. Jonathan Friedman and Kajsa Ekholm Friedman (2013) also point out the differentiated meanings of globalization by class: for (upper) middle-class cosmopolitans who shop for various opportunities globally and for the working class who face global competition for jobs.

Despite such criticisms, the notion of the global continues to be celebrated in various arenas, from education, where nurturing "global competence" and creating "global citizens" remain its goals, to job markets that value such "global" experiences (Urciuoli, 2016, 2018). "Global competence" includes many features without consensus among researchers, though some common threads according to Hunter et al. (2006) are: (1) various types of international knowledge on topics such as world history and events; (2) skills like awareness of and adaptability to diverse cultures; (3) beliefs and attitudes such as openness to difference; and (4) competencies within specific disciplines such as the ability to collaborate across cultures. The oft-quoted definition by Richard Lambert (1994) lists: (1) knowledge of globally cross-cutting issues like environment, energy, food, and human rights, or area-specific deep knowledge; (2) empathy, "the ability of an individual to psychologically put her or himself into another person's shoes" (1994, p. 15), which entails a progression from ethnocentric to "ethnorelativistic" (1994, p. 16) standpoints via the stages of acceptance, adaptation, and integration; (3) approval or favoring of things abroad; (4) foreign language competency; and (5) task performance in international arenas. "Global imagination" that allows one to envision a plurality of the imagined world and tolerance for ambiguity are sometimes listed as part of global competence as well (Brockington & Wiedenhoeft, 2009; Rizvi, 2000; Skelly, 2009; Streitwieser, 2009). A related notion, "global citizen," includes similar attributes such as "(1) global knowledge; (2) understanding the interconnectedness of the world in which we live; (3) intercultural competence, or the ability to relate successfully with those from other cultures; and (4) engagement on the local and global level around issues that impact humanity" (Deardorff, 2009, p. 348).

I have critiqued elsewhere (Doerr, 2018, 2020) regarding who decides what constitutes global competence, which reflects relations of power and celebrates specific types of knowledge and skills, often reflecting the aforementioned Regimes of Mobility. For example, the same travel experience can be celebrated as a "global experience" if done by White middle-class students studying abroad, whereas ignored if done by non-White immigrant students, who may travel to their ancestral homeland that is still a "different culture" for them (Doerr, 2018, 2020; Doerr et al., 2020; Moll et al., 2005). Michael Woolf (2010) calls the claim to transform students into global citizens through study abroad programs an unachievable hyperbole that masks and distorts the tangible benefits of study abroad and supports the idea of global citizens as a new privileged class. Talya Zemach-Bersin (2009, p. 317) critiques the study abroad advertisements that posit global citizenship as something to be purchased through the study abroad experience, promoted in conjunction with themes of personal advancement and consumerism.

Despite such critiques, the notions of global competence and global citizenship prevail as something positive, which employers seek in their employees, pushing individuals to seek them as a way to strengthen their resume. Chapters in this volume illustrated this through cases where individuals sought to brand themselves as a global neoliberal subject (Schedel) or where individuals who lacked cultural capital or could not gain "global experience" earlier sought it through employment, even if the jobs were unpaid (Jang) or discriminatory (Choi) or included unpaid portions in off-duty hours (McDaid & Sunyol). In these contexts, the valorization of the global allowed the global experience to become the wages that compensated for unpaid—volunteer—work.

The Discourse of Immersion

The second discourse that backgrounds the Wages of Global Experience is that of immersion, which is linked to the above discourse of globalization in that "global competence" is seen to be gained through immersion. In the field of study abroad, immersion—"live like a local"—is considered the best way to gain "global competence" and to become "global citizens" by understanding "another culture," being sympathetic to people of other cultural backgrounds, gaining proficiency in another language, and learning to navigate an unknown environment (Brockington & Wiedenhoeft, 2009; Currier et al., 2009; Cushner, 2009; Kinginger, 2008).

The discourse of immersion suggests students leave their "comfort zone"—where one stays connected to friends and family back home, spends time with fellow students from their home country, or buried in the world of the Internet, TV, and computers (Hovey & Weinberg, 2009; Loflin, 2007; Oxford, 2005; Williamson, 2004)—because studying abroad is a zero-sum game where the "comfort zone" and host society vie for a student's time, interest, and commitment. Comfort zones, it is assumed, consume time, drain mental energy, and interrupt language learning, rather than work as a buffer for those who would otherwise feel anxious about leaving their home country.

The discourse of immersion also calls for proactive engagement with people and activities in the host society through, for example, staying with a host family, making local friends, talking with strangers, and even getting lost (Cohen et al., 2005; Goldoni, 2013; Jackson, 2009; Loflin, 2007; Raymond & Hall, 2008; Vande Berg et al., 2009; Williamson 2004). Some researchers consider such immersion not enough unless accompanied by more intentional activities, such as ethnographic projects (Goldoni, 2013; Ogden, 2006; Roberts et al., 2001); getting help from a well-trained "cultural mentor" to reflect on, hypothesize about, and actively test cultural concepts and skills (Vande Berg, 2009, p. 23); and using reflective writing (Chen, 2002) to increase opportunities for reflection. Doing volunteer work, as mentioned in the chapters of this volume, is often thought of as a way to increase one's engagement with local people and thus increase immersion (Bringle et al., 2011; Plater et al., 2009).

I have critiqued elsewhere (Doerr, 2013) the ways that the discourse of immersion creates a blanket hierarchy of the learning experience—the out-of-class component over classroom learning (Chen, 2002; Laubscher, 1994; Peterson, 2002); staying with a host family over staying in a hostel or dormitory with fellow compatriot students (Chieffo & Griffiths, 2004; Gutel, 2007; Hovey & Weinberg, 2009); long-term over short-term stays (Currier et al., 2009; Deardorff, 2009); direct enrollment over island-type programs (for a critique, see Woolf, 2007); and spending time with the locals over fellow compatriot students (Deardorff, 2009; Hovey & Weinberg, 2009)—without considering the possible positive effects of the unvalued practices above such as comparing experiences with fellow compatriot students who share the experience of both host and home society (Woolf, 2007). I have also critiqued (Doerr, 2013) that the discourse of immersion suggests the host and home societies to be internally homogeneous, with the former lacking global connections.

Nonetheless, the immersion experience continues to be valorized not only in study abroad but also in general learning contexts involving cross-cultural experience. This understanding is the backdrop, for example, for the views of individuals discussed in Kyoko Motobayashi's chapter. Her research participants viewed the immersion experience during their off-duty time volunteering in Latin America as an important factor in their decision of where they would go to teach Japanese. Also, Peruvian child vendors who prefer learning English outside the classroom through conversing with voluntourists from the U.S., as Sinervo described, suggest the value of immersion as creating connections among those involved, which becomes important in creating long-term mentor/guardian-type relationships.

The positioning of immersion—merely existing and experiencing mundane life—as something meaningful and thus worth paying a lot of money for, as in studying abroad focused on immersion, underlies individuals putting up with unpaid labor: if immersion is something worth paying for, getting free immersion can balance the free labor they provide. The chapters by Jang, Choi, and McDaid and Sunyol illustrated this, suggesting the "Wages of Immersion" that legitimize the "Wages of Global Experience."

This is not to say such immersion experience or global experience is meaningless (see Palmer, 2015). For example, Choi's chapter explained the way South Korean working holiday makers who are discriminated against as a minority in Australia came to realize how they themselves discriminated against minority migrants in their homeland of South Korea. However, if such immersion experiences are valued to the degree that one would pay money for them, and thus getting them "for free" is seen as a wage (i.e., "Wages of Immersion") that compensates the free labor one provides, it becomes problematic.

The Discourse of Volunteering as Receiving

The third discourse that backgrounds the Wages of Global Experience is the notion that volunteering is rewarding on top of being a learning experience. Some argue that volunteering is not merely about giving, as volunteers do gain something in return. Such returns include distinguishing themselves from other social classes (Mowforth & Munt, 2009); gaining a competitive edge in the educational market (Gray & Campbell, 2007; Heath, 2007; Munt, 1994); building personal qualities like strength of character, adaptability, sensitivity, and worldliness (Gray & Campbell, 2007; Munt, 1994); constructing themselves as "White," "moral," "bourgeois," and "superior" (Angod, 2015; Heron, 2007; Tiessen & Huish, 2013); experiencing a rite of passage in maturing (cited by McDaid & Sunyol); or feeling good about themselves (Jakubiak, 2012).

Volunteering as a learning experience is epitomized in the concept of service-learning, which developed in the 1990s, with the rise of volunteer work via: (1) a shift away from the Cold War Left/Right grand narrative and collective politics toward an individual-centered "life politics," focused on individual morality and sense of self; (2) a post-Fordist shift from the collective politics of production and social classes to consumption and an individual sense of self (Butcher & Smith, 2010); (3) the spread of neoliberalist reforms that let the state withdraw from social services, leaving non-governmental organizations (NGOs) with volunteers to fill the gap (Conran, 2011); and (4) the emergence of the "ever-increasing 'guilt- conscious' society" through the awareness of wealth disparity (Callanan & Thomas, 2005, p. 183). This was a shift away from post-World War II programs like Voluntary Service Overseas and the U.S. Peace Corps, which are (1) altruistic, fighting poverty under the assumption that Westernization/modernization is desirable; (2) political, designed to promote the image of Western altruism and reduce the threat of communist influence; and (3) carried out by skilled people (Butcher & Smith, 2010).

The attention to service-learning as part of education at school in the U.S. was initiated by Benjamin Barber who viewed, following John Dewey, service as a duty of responsible citizens of a democratic society and thus called to mandate service-learning in schools (Barber, 1994; also see Rhoads & Neururer, 1998; Saltmarsh, 1996; Taylor, 2002). Service-learning was also seen as linking community service to the sphere of civic education with the goal of developing "a sense of personal responsibility and empathy for the larger community, particularly the community beyond the confines of the college campus" (Rhoads & Neururer, 1998, p. 101).

The crossing of economic, social, and cultural boundaries through service-learning came to be seen as an important learning experience called border pedagogy. Assuming service as White middle-class students serving in underprivileged minority communities, service-learning is seen as "deliberate encounters with diverse cultures" (Rhoads & Neururer,

1998, p. 112) where students cross three kinds of boundaries: (1) physical boundaries, as students leave the classroom to enter communities; (2) socially constructed boundaries such as race, class, age, and religion; and (3) epistemological and pedagogical boundaries, whose transgression creates space for new ways of knowing and learning (Taylor, 2002; also see Green, 2001). College is usually seen as a "bubble" of privilege in relation to the off-campus "real world" that is "in need," to which students cross the boundary to serve and learn through gaining different perspectives of the world (LaDousa, 2014).

There were also institutional pushes in stimulating interest in community service. For example, the National Community Service Act of 1990 offered student loan deferments to borrowers who performed volunteer services. The state of Maryland made 75 hours of service-learning a requirement for high school graduation (Taylor, 2002). College alternative break trips flourished with the support of newly established organizations such as Campus Compact, Campus Outreach Opportunity League, the National Society for Experiential Learning, Break Away, and the Partnership for Service-Learning (Rhoads & Neururer, 1998). Neoliberalist transformation led higher education institutions to operate as businesses, pushing students' "experiences"—especially those of studying abroad and doing service-learning—and the soft skills they supposedly gain from them to be recast as objectified items to help brand the university as well as mark the students' employability (Urciuoli, 2016, 2018).

Such understandings of volunteering as not just giving but receiving—not only for self-advancement but also learning and more—background the perception of volunteer work discussed in the chapters of this volume, leading my argument of the Wages of Global Experience. Jang's chapter's description of South Korean study abroad students' preferring unpaid volunteer work because it gives them "work experience in the West" on their resume over paid menial work with no linguistic/work experience makes sense in this light. Schedel's chapter on a man with a legally undocumented status working as a voluntourist as a way to secure living arrangements because volunteer work does not require a work visa is another example. McDaid and Sunyol's chapter showed a similar case, where low wage/underpayment for their work was justified by the workers or their parents as what they are receiving—immersion in the Irish life, the Wages of Global Experience.

Effects on the Discourses of Globalization, Immersion, and Volunteering as Receiving

Perceiving voluntourism as compensated by the Wages of Global Experience allows us to revisit these discourses and recognize their effects. As Du Bois and then Roediger used the economic metaphor and showed how the Wages of Whiteness both kept the actual wage low and nurtured racism among the White working class, I argue that the Wages of Global Experience both kept the actual wage low (including no wage) and nurtured globalism, the valorization of the global connections and experiences. Like the Wages of Whiteness worked to oppress African Americans via White working-class racism while oppressing White working-class people themselves, the Wages of Global Experience work to devalue "non-global" experiences and those who only have such "non-global" experience (that pushed these volunteers to seek the global experience to start with) and exploit volunteers by not paying them. That is, with this theorization, we can see the detrimental effects of globalism and its valorization of global experience.

The discourse of globalization creates the value of "global experience" that gets used in a way that can hide exploitation because it suggests value that can be exchanged for labor, where there is little extra labor done by those who are "creating" the global experience for them (i.e., they are just living their own mundane lives which creates the value only because it is done in a different setting from the volunteers' homeland, hence "global"). This is similar to the situation of the commodification of immersion discussed below. The difference is that global experience is value-added because of the valorization of the global, pointing to another negative effect of globalism.

The discourse of immersion makes the immersion experience itself a commodity, as is seen in study abroad. The people who provide immersion experiences are not supposed to do any labor, because the value is in sharing the mundane, "usual," "the way it is," experience with the visitors/study abroad students/voluntourists, and not in any special program that they have created. Immersion's commodification then creates an imbalance. When such immersion is freely given, they become the Wages of Global Experience that appear to compensate for the lack of payment, as in the case of voluntourism discussed in this volume. That is, the commodification of immersion is problematic not only for its introducing consumerism in education (Zemach-Bersin, 2009) but also for setting up a situation that is conducive to exploitation where immersion occurs. Merely living in the destination sounds like a "good deal," as if one is receiving a "free commodity," although nobody is getting paid for producing said commodity (hence the intermediaries reap the profit, as Jang showed).

The discourse of volunteering as receiving is, in one sense, a humbling expression that can help reduce the helper-helped hierarchy as giver-receiver relations (Henry & Breyfogle, 2006). However, such a perception can be used to justify the unpaid aspect of volunteer work because the employer can claim that volunteers are getting something out of it, even if it is not monetary. This is conducive to accepting the Wages of Global Experience. The notion of the Wages of Global Experience then problematizes this apparently humble perception of "volunteering as receiving," pushing us to explore other ways of framing the volunteer experience.

Language Learning, Standardization, and Unit Thinking: Reinterpreting Ideologies of Native Speakers and Immersion

Jakubiak's chapter problematized Native Speakerism, which was further critiqued in Bernstein's as the measure to correct it—TEFL—fell short through its commodification. The chapters by Sinervo, Motobayashi, Schedel, Jang, Choi, and McDaid and Sunyol pointed to the assumed value of immersion that backgrounds their arguments. These discussions inspired me to think of my own interest in language politics: language standardization through language teaching. In this section, in light of the issue of standardization, I reinterpret these chapters' discussions to suggest a possible alternative to language teaching that can defy the standardization of the target language in due course. That is, I show below how a critique in one context—such as the native speaker ideology—can be reframed in another context (e.g., language-motivated voluntourism) for a better effect, in due course modifying the notion of the "native speaker" itself.

Language standardization is an effect of relations of power, where the speech of the dominant group gets imposed onto others. Though the process occurred in pre-modern times as well, as reported about Elio Antonio de Nebrija creating the standardized grammar of Castilian and presenting it to Queen Isabella of Spain (Illich, 1981; Train, 2009), it became prevalent with the birth of nation-states and the ideology of one nation, one people, one language. Johann Gottfried von Herder's formulation of "Volk" as an internally homogeneous group speaking a homogeneous language in late eighteenth-century Germany served as the model *for* the nation-state (Balibar, 1994; Bauman & Briggs, 2000). It then made the diversity in speech be homogenized through standardization processes via schooling among other institutions (Bourdieu, 1991).

This promoted the notion of language as a bounded, internally homogeneous unit—what I called "unit thinking" elsewhere (Doerr, 2022)—that paradoxically supported standardization efforts: paradoxical because if the unit of language is internally homogeneous, there should be no need for active standardization. I have discussed this in a wider context of modernity where various things—people, culture, and language, among others—are seen as discrete bounded units with homogeneous insides (i.e., "unit thinking") that align with the nation-state ideology. Language standardization is one example of an institutionalized action based on unit thinking to fit the heterogeneous linguistic practice into normalized standard form (Doerr 2022, 2023).

Although researchers critiqued the standardizing effects of language education, efforts to challenge its complicity have been restricted to an explicit critique of the inherent power relations behind standardization processes rather than incorporating pedagogy that would defy such standardization processes. That is, while critiquing the oppressive process of having to change one's speech or the hierarchies among linguistic varieties created by the standardization processes, the teaching material remained the standard form. This practice was often supported by the claim that students needed to learn the most "desirable" form, which is the standard (Mori et al., 2020).

I critiqued this approach elsewhere (Doerr, 2023, in preparation) as half-baked, if not hypocrisy, and have suggested alternative pedagogical approaches. One of the suggestions I made that is relevant here is what I called "neo-immersion," based on the pedagogic approach of the Master-Apprentice Language Learning Program developed in the context of language revitalization efforts among Native American groups in California (Hinton et al., 2002). Though this method was developed because of a lack of resources (Hammine, 2022, forthcoming), its approach can be applied to other contexts, such as foreign language teaching, as a way to avoid standardization processes in language teaching. Below I introduce the notion of neo-immersion by describing the Master-Apprentice Language Learning Program, suggesting a new notion to replace the "native speaker" concept—"comfortable speakers (formerly known as native speakers)"—and connect them to this volume's discussions.

The Master-Apprentice Language Learning Program and Neo-Immersion

The Master-Apprentice Language Learning Program is based on the theory of learning language through immersion that holds that "adults can learn language informally, through listening, speaking, and eliciting language from a native speaker, and mainly by doing activities together in which the language is being used" (Hinton et al., 2002, p. 7). Its main principles are (1) to use the target language only; (2) the Apprentice/learner is as active as the Master/teacher in deciding what will be learned; (3) a focus on oral skills rather than written; (4) learning in daily situations of carrying out tasks (e.g., cooking, gardening); and (5) the activity and the non-verbal communication provide the contextual clues (Hinton, 2001). Based on the Total Physical Response or TPR (Asher, 1966) and drawing on language immersion programs like Hawaiian Aha Pūnana Leo, conversational competence models, and linguistic elicitation techniques, the Master-Apprentice Language Learning Program is a one-on-one teaching practice between the "master" (teacher/speaker) and the "apprentice" (language learner), often the elder who grew up speaking the language and a younger person learning the language (Hinton, 2001).

Based on the suggestion of Julian Lang (a Karuk speaker), the Master-Apprentice program was developed by the Native California Network in 1992. With the initial design by Leanne Hinton, Nancy Richardson, Mary Bates Abbott, and others and refined by the California Foreign Language Project, the program was administered for seven years by the Advocates for Indigenous California Language Survival (AICLS). A Master-Apprentice team applied and was selected by the AICLS board based on the criteria of the Master's fluency, the Apprentice's interest in learning and teaching the language, and how close the Master and Apprentice are (they apply as a team). The team worked together for up to three years, after which they could continue on their own. Each member was given a stipend of $3000 for 360 hours of language immersion work. The Apprentice was paid because they needed to take time off from their paid work to do this project. The Apprentice kept a log of the learning activities and sent its 40-hour worth of work to the coordinator, after which the check for those 40 hours was sent. Unlike immersion in the study abroad context, the Master-Apprentice Program's immersion is more structured, with each session planned ahead of time.

Each team underwent an assessment at the beginning and end of each year, carried out in two parts. In the first part, questions in English were given to the Master, who then translated them and asked the Apprentice those questions. The Master kept asking questions until the Apprentice did not understand the question. While the Master evaluated the grammatical accuracy of the Apprentice's answers, the Assessors evaluated only the seeming fluency—whether the answers were in long sentences and flowing without stopping often—not grammatical accuracy (the Assessors usually do not understand the language). This aspect is important and will be further discussed in the next section. In the second part, the Apprentice chose a picture of their Native American group doing activities out of a collection and talked about the picture, sometimes with a prompt from the Master. The Master translated the Apprentice's words and the Assessors wrote down the fluency and length of the description. Assessments were videotaped and compared to later ones (Hinton, 2001).

The benefits of this program, Hinton reports, are, first of all, that the Apprentice usually became conversationally proficient after three years and was ready to teach others. Also, this program can involve elders in positive ways by boosting their self-esteem, in contrast to school-based programs that may alienate them. This Master-Apprentice model has spread to various indigenous groups (Hammine, 2022; Hinton, 2001).

While this Master-Apprentice Language Learning Program provides a useful model for revitalizing marginalized languages, I here suggest applying this for evading unit thinking in language education. Because of its design of one-on-one teaching relationships where the learner (the Apprentice) learns from one teacher (the Master), mimicking their specific speech—idiolect—rather than a standardized version of the language, it can work as a model to evade standardization in language education. I call this neo-immersion (Doerr, 2023, in preparation).

The neo-immersion approach is to treat individuals as masters in this Master-Apprentice model. Individuals are thus seen as speaking their own fluid idiolect, which the learner/Apprentice would learn. The more masters the learner has, the wider the vocabulary and the more ways of speaking the learner gains.

Here, although appreciating its main model coming from the Master-Apprentice Language Learning Program, I modify the name of this method to neo-immersion because of its new focus on defying standardization (Master-Apprentice Program may support standardization of the language). More importantly, I modify the naming of the teacher from the "Master" to merely "language partner." This renaming is especially important in the context of English-language voluntourism as I will further discuss: the hegemonic spread of English as the global language, which Robert Philipson (1992) called "linguistic imperialism." The original formulation of the Master-Apprentice Language Learning Program was indigenous language revitalization, which positioned the elders as the language teachers. In that context, showing respect to the elders and to the carriers of heritage through the notion of "master" made sense. Yet, when applying this method in different contexts, including English-language voluntourism, the culture surrounding necessitates the renaming of this process.

Comfortable Speakers (Formerly Known as Native Speakers)

This neo-immersion approach also leads to the reformulation of the "native speaker" concept based on the Master-Apprentice Language Learning Program's evaluation process where the fluency and confidence are the source of evaluation. Drawing on this approach, I modified the problematic notion of the "native speaker" with a new name "comfortable speaker (formally known as the native speaker)" (Doerr, 2023).

This new notion of "comfortable speaker" allows us to move away from the three problematic aspects of the "native speaker" ideology as identified by Alistair Pennycook (1994). The first aspect is the close correspondence between holding the citizenship of a nation-state and being a native speaker of the national language of that nation-state (Pennycook, 1994). Here, there is an assumption that an individual speaks one language, reflecting the unit thinking of language and its erasure of heterogeneity of language practices in relations of power. For example, Tove Skutnabb-Kangas and Robert Philipson (1989) showed four possible definitions of "mother tongue" (or "native language"): (1) the language one learns first; (2) the language one knows best; (3) the language one uses the most; and (4) the language one identifies with the most. In this light, it does not make sense to identify one language as one's "native language" and, if we see it from the reverse side, identifying who is a "native speaker." Choosing the second and third definitions of "mother tongue" here, I suggest replacing the notion of the "native speaker" with a new notion of "comfortable speaker (formerly known as the native speaker)."

The second aspect of the native speaker ideology is the notion that language is a homogeneous and fixed system with a homogeneous speech community, which allows "a rigid and clear distinction between being a native speaker and not being so" (Pennycook, 1994, p. 176). John Milroy and Lesley Milroy (1991) argued that the standardization process never ends because language always changes: the only completely standardized language is a dead language. Though "native speakers" are the target model language speakers to be emulated in the foreign/second language classrooms, they themselves are the target of standardization within their own education system (i.e., language arts classes), although they are never totally standardized (Doerr, 2009). These "native speakers" thus continue to speak their own fluid idiolect. My notion of "comfortable speaker" includes this aspect of fluid heterogeneity of individual speakers and their speech. It also incorporates the idea that one's utterances are formed contextually in relation to the interlocutor's speech, something that emerges in conversation rather than is "owned" by the speaker (see Vološinov, 1973).

The third aspect of the native speaker ideology is the idea that being a "native speaker" automatically bestows one with a high level of competence in all domains of one's first language, implying that the "native speaker" has "a complete and possibly innate competence in the language" (Pennycook, 1994, p. 175). Here, what is important is not just that "native speakers" do make "grammatical mistakes," but also that mistakes in language are a fluid thing that cannot be used as a yardstick. Considering that language constantly changes and that changes are led by the "native speakers" who start using the language differently—identifying "mistakes" by "native speakers" is a numbers game: if enough "native speakers" use that form, it becomes "correct." For that reason, talking about speakers "who do not make mistakes" is tautological because their speech defines "grammatical correctness." Therefore, it does not make sense to discuss who, whether native speaker or not, is innately competent in any specific language. Moving away from such yardsticks of "correctness," I use the notion of the comfort level of speaking—hence "comfortable speaker"—and suggest replacing the notion of the native speaker.

Rethinking English-Language Voluntourism and Immersion

This reformulation of the notion of "native speaker" to the new notion of "comfortable speaker" allows us to rethink the English-language voluntourists' potential. If we view them as "comfortable speakers," then English-language voluntourists' presence in places they work can be formulated into neo-immersion in reverse (rather than voluntourist immersing in the host society), where the people in the destination can learn English from the voluntourists on a one-on-one basis. If the training in TEFL can be seen as a form of standardization, with the added problem of commodification as Bernstein analyzed, voluntourism may be better off without such training, possibly replacing that with the aforementioned training in the Master-Apprentice Language Learning Program that is borrowed by the neo-immersion framework.

It is worth noting, however, that though language standardization processes may be halted, the problematic nature of English-language voluntourism that Jakubiak suggested can continue to prevail even if we reformulate it as neo-immersion in reverse and rename the "Master" derived from the Master-Apprentice Language Learning Program into a mere "language partner."

What this volume suggests as a whole, with chapters covering diverse contexts and topics, nonetheless, may give us new directions. The mutual exchange, if not reciprocity, between the labor of the volunteers and what they gain from it, be it the "Wages of Global Experience" or free accommodations, points to a two-way process. Sinervo's chapter suggested the voluntourists and Peruvian child vendors mutually teaching English (for the children's future career) and Spanish (for forging connections) to each other. Motobayashi's chapter showed the volunteer Japanese language teachers being also students learning the local language as an important aspect of their experience. Free or underpaid labor of study abroad students (Jang, Schedel), working holidaymakers (Choi), and language assistants (McDaid and Sunyol) being exchanged for (though unfairly) something they appreciated suggests mutual processes that can be incorporated in framing language exchange. Although I continue to problematize them as unbalanced and forced, as their justification serving as a way to come to terms with unfavorable conditions, I here use them as a possible model if transferred to the language learning situation when the volunteer experience involves language teaching.

That is, such language-motivated voluntourism can be reformulated (not only at the level of perception but also at the level of practice) as mutual processes of volunteers and volunteered being language partners who are "comfortable speakers (formerly known as native speakers)" of their fluid language, teaching each other their own idiolect, skipping the standard language of textbooks. Both sides can be trained in the Master-Apprentice Language Learning Program model to make the learning process more structured without standardizing the content of the language: neo-immersion. That way, we can talk about volunteers gaining something but in a more formalized way that evades creating hierarchy between the volunteers and the volunteered, the helper and the helped (see Henry & Breyfogle, 2006).

Standardization is built on and perpetuates relations of power between those whose daily speech approximates the standard language (i.e., the dominant group in the society) and those whose daily speech does not (i.e., marginalized groups). Evading such standardization process through language learning thus challenges relations of power in linguistic practices and beyond. Such a new framework—neo-immersion—can reformulate not only linguistic practices in societies where volunteers are from as well as where they work but also the framework of thinking about language and language education in general.

Wordplay

This may all mount to wordplay, albeit an important one. The same thing can be called differently depending on how it is perceived, which leads to different actions. Chapters in this volume revealed this importance of naming. "Voluntourists" teaching English can be seen as neo-colonialists (Jakubiak; Bernstein; Sinervo) in the process of "linguistic imperialism" (Phillipson, 1992). Unpaid "internships" (Jang), "volunteers," and even homestays where students babysit host siblings can well be called "slave labor." That is, the "volunteer" label can be used as a way to justify no/low wage as illustrated in chapters by Jang, Choi, McDaid and Sunyol, as discussed. "Volunteering" or underpaid working holiday experiences can also be considered "free study abroad" to learn local languages (Motobayashi; Choi) or "work-visa evasion" (Schedel). Naming can make a practice seem legitimate or less so. And therein lies the importance of the work researchers do: changing the perceptions, changing the practices.

Connecting diverse practices that can be called language-motivated voluntourism and tracing various economic, social, and cultural mechanisms that are involved in shaping their practices, the chapters in this volume illuminated the effects of naming, which is what theorization is. This volume leads us in many directions, depending on what aspects we pay attention to and connect our own work and interests to, like I did with the notion of the Wages of Global Experience, language standardization, and the native speaker concept. This embodies the beauty of an edited volume that relates diverse works that show us this important wordplay—what they make us see and what they hide—creating sparks for us to chase and investigate further.

References

Angod, L. (2015). *Behind and beyond the ivy (in italics). [Unpublished doctoral dissertation]*. University of Toronto. https://tspace.library.utoronto.ca/handle/1807/69729

Asher, J. J. (1966). The learning strategy of the Total Physical Response: A review. *The Modern Language Journal, 50*(2), 79–84. https://doi.org/10.2307/1127119

Balibar, E. (1994). *Masses, classes, ideas: Studies on politics and philosophy before and after Marx*. Routledge.

Barber, B. R. (1994). A proposal for mandatory citizen education and community service. *Michigan Journal of Community Service Learning, 1*, 86–93. http://hdl.handle.net/2027/spo.3239521.0001.110

Bauman, R., & Briggs, C. (2000). Language philosophy as language ideology: John Locke and Johann Gottfried herder. In P. V. Kroskrity (Ed.), *Regimes of language: Ideologies, polities, and identities* (pp. 139–204). School of American Research Press.

Bourdieu, P. (1991). *Language and symbolic power*. Harvard University Press.

Bringle, R. G., Hatcher, J. A., & Jones, S. G. (2011). *International service learning: Conceptual frameworks and research*. Stylus Publishing.

Brockington, J. L., & Wiedenhoeft, M. D. (2009). The liberal arts and global citizenship: Fostering intercultural engagement through integrative experiences and structured reflection. In R. Lewin (Ed.), *The handbook of practice and research in study abroad: Higher education and the quest for global citizenship* (pp. 117–132). Routledge.

Butcher, J., & Smith, P. (2010). "Making a difference": Volunteer tourism and development. *Tourism Recreation Research, 35*(1), 27–36. https://doi.org/10.1080/02508281.2010.11081616

Callanan, M., & Thomas, S. (2005). Volunteer tourism: Deconstructing volunteer activities within a dynamic environment. In M. Novelli (Ed.), *Niche tourism: Contemporary issues, trends and cases* (pp. 183–200). Elsevier. https://doi.org/10.1016/B978-0-7506-6133-1.50025-1

Chen, L. (2002). Writing to host nationals as cross-cultural collaborative learning in study abroad. *Frontiers: The Interdisciplinary Journal of Study Abroad, 8*, 143–164. https://doi.org/10.36366/frontiers.v8i1.98

Chieffo, L., & Griffiths, L. (2004). Large-scale assessment of student attitudes after a short-term study abroad program. *Frontiers: The Interdisciplinary Journal of Study Abroad, 10*, 165–177. https://doi.org/10.36366/frontiers.v10i1.140

Cohen, A. D., Paige, R. M., Shively, R. L., Emert, H. A., & Hoff, J. G. (2005). *Maximizing study abroad through language and culture strategies: Research on students, study abroad program professionals, and language instructors*. Center for Advanced Research on Language Acquisition.

Conran, M. (2011). They really love me! Intimacy in volunteer tourism. *Annals of Tourism Research, 38*(4), 1454–1473. https://doi.org/10.1016/j.annals.2011.03.014

Currier, C., Lucas, J., & Arnault, D. S. (2009). Study abroad and nursing: From cultural to global competence. In R. Lewin (Ed.), *The handbook of practice and research in study abroad: Higher education and the quest for global citizenship* (pp. 133–150). Routledge.

Cushner, K. (2009). The role of study abroad in preparing globally responsible teachers. In R. Lewin (Ed.), *The handbook of practice and research in study abroad: Higher education and the quest for global citizenship* (pp. 151–169). Routledge.

Deardorff, D. (2009). Understanding the challenges of assessing global citizenship. In R. Lewin (Ed.), *The handbook of practice and research in study abroad: Higher education and the quest for global citizenship* (pp. 346–364). Routledge.

Doerr, N. M. (2009). Investigating "native speaker effects": Toward a new model of analyzing "native speaker" ideologies. In N. M. Doerr (Ed.), *The native speaker concept: Ethnographic investigations of "native speaker effects"* (pp. 15–46). Mouton de Gruyter.

Doerr, N. M. (2013). Do "global citizens" need the parochial cultural other? Discourses of study abroad and learning by doing. *Compare, 43*(2), 224–243. https://doi.org/10.1080/03057925.2012.701852

Doerr, N. M. (2018). *Transforming study abroad: A handbook*. Berghahn Books. https://doi.org/10.3167/9781789201154

Doerr, N. M. (2020). "Global competence" of minority immigrant students: Hierarchy of experience and ideology of global competence in study abroad. *Discourse: Studies in the Cultural Politics of Education, 41*(1), 83–97.

Doerr, N. M. (2022). *Fairies, ghosts, and Santa Claus: Tinted glasses, fetishes, and the politics of seeing*. Berghahn Books. https://doi.org/10.3167/9781800736870

Doerr, N. M. (2023). *"Neo-immersion" and "neo-test-prep" pedagogies: Beyond the standard and unit-thinking in language*. Paper presented at the annual meeting of the American Association for Applied Linguistics, Portland, OR.

Doerr, N. M., Puente, D., & Kamiyoshi, U. (2020). Global citizenship, identity and intercultural competence: Student identities in education abroad. In A. C. Ogden, B. Streitwieser, & C. Van Mol (Eds.), *Education abroad: Bridging scholarship and practice* (pp. 119–134). Routledge.

Du Bois, W. E. B. (2021/1935). *Black reconstruction: An essay toward a history of the part which black folk played in the attempt to reconstruct democracy in America, 1860–1880*. Library of America.

Friedman, J., & Friedman, K. E. (2013). Globalization as a discourse of hegemonic crisis: A global systemic analysis. *American Ethnologist, 40*(2), 244–257. https://doi.org/10.1111/amet.12017

Freire, P. (1997/1970). *Pedagogy of the oppressed*. Continuum.

Glick Schiller, N., & Salazar, N. B. (2013). Regimes of mobility across the globe. *Journal of Ethnic and Migration Studies, 39*(2), 183–200. https://doi.org/10.1080/1369183X.2013.723253

Goldoni, F. (2013). Students' immersion experiences in study abroad. *Foreign Language Annals, 46*(3), 359–376. https://doi.org/10.1111/flan.12047

Gray, N. J., & Campbell, L. M. (2007). A decommodified experience? Exploring aesthetic, economic and ethical values for volunteer ecotourism in Costa Rica. *Journal of Sustainable Tourism, 1*(5), 463–482. https://doi.org/10.2167/jost725.0

Green, A. E. (2001). "But you aren't White": Racial perceptions and service-learning. *Michigan Journal of Community Service Learning, 8*(1), 18–26.

Gutel, H. (2007). The home stay: A gendered perspective. *Frontiers: The Interdisciplinary Journal of Study Abroad, 15*, 173–188. https://doi.org/10.36366/frontiers.v15i1.225

Hammine, M. (2022). Hear our voice: New speakers of *ryukyuan* language—Negotiation, construction, and change of identities. In M. Mielick, R. Kubota, & L. Lawrence (Eds.), *Discourses of identity: Language learning, teaching, and reclamation perspectives in Japan* (pp. 117–137). Springer.

Heath, S. (2007). Widening the gap: Pre-university gap years and the "economy of experience". *British Journal of Sociology of Education, 28*, 89–103.
Henry, S. E., & Breyfogle, M. L. (2006). Toward a new framework of "server" and "served": De(and re)constructing reciprocity in service-learning pedagogy. *International Journal of Teaching and Learning in Higher Education, 18*(1), 27–35.
Heron, B. (2007). *Desire for development*. Wilfred Laurier University Press.
Hinton, L. (2001). The master-apprentice language learning program. In H. Hinton & K. Hale (Eds.), *The green book of language revitalization in practice* (pp. 217–226). Brill. https://doi.org/10.1163/9789004261723_018
Hinton, L., Vera, M., & Steele, N. (2002). *How to keep your language alive: A commonsense approach to one-on-one language learning*. Heyday.
Hovey, R., & Weinberg, A. (2009). Global learning and the making of citizen diplomats. In R. Lewin (Ed.), *The handbook of practice and research in study abroad: Higher education and the quest for global citizenship* (pp. 33–48). Routledge.
Hunter, B., White, G. P., & Godbey, G. C. (2006). What does it mean to be globally competent? *Journal of Studies in International Education, 10*(3), 267–285. https://doi.org/10.1177/1028315306286930
Illich, I. (1981). *Shadow work*. Marion Boyars.
Jackson, J. (2009). Intercultural learning on short-term sojourn. *Intercultural Education, 20*(1–2), 59–71. https://doi.org/10.1080/14675980903370870
Jakubiak, C. (2012). "English for the global": Discourses in/of English-language voluntourism. *International Journal of Qualitative Studies in Education, 25*(4), 435–451. https://doi.org/10.1080/09518398.2012.673029
Kinginger, C. (2008). Language learning in study abroad: Case studies of Americans in France. *Modern Language Journal, 92*(S1), 1–124. https://doi.org/10.1111/j.1540-4781.2008.00821.x
LaDousa, C. (2014). "Everyone's got room to grow": A discourse analysis of service-learning rhetoric in higher education. *Learning and Teaching, 6*(2), 33–52.
Lambert, R. D. (1994). Parsing the concept of global competence. In R. Lambert (Ed.), *Educational exchange and global competence* (pp. 11–24). Council on International Educational Exchange.
Laubscher, M. R. (1994). *Encounters with difference: Student perceptions of the role of out-of-class experiences in education abroad*. Greenwood Press.
Loflin, S. E. (2007). *Adventures abroad: The student's guide to studying overseas*. Kaplan Publishing.

Milroy, J. V., & Milroy, L. V. (1991). *Authority in language: Investigating language prescription and standardization* (2nd ed.). Routledge.

Moll, L., Amanti, C., Neff, D., & Gonzalez, N. (2005). Funds of knowledge for teaching: Using a qualitative approach to connect homes and classrooms. In N. Gonzalez, L. C. Moll, & C. Amanti (Eds.), *Funds of knowledge: Theorizing practices in households, communities, and classrooms* (pp. 71–87). Routledge.

Mori, J., Hasegawa, A., Park, J., & Suzuki, K. (2020). On goals of language education and teacher diversity beliefs and experiences of Japanese-language educators in North America: Exploring who we are and who we aspire to be diversity, inclusion, and professionalism in Japanese language education. *Japanese Language & Literature, 54*(2), 267–304. https://doi.org/10.5195/jll.2020.131

Mowforth, M., & Munt, I. (2009). *Tourism and sustainability: Development, globalization and new tourism in the third world.* Routledge.

Munt, I. (1994). Eco-tourism or ego-tourism? *Race and Class, 36*(1), 49–60. https://doi.org/10.1177/030639689403600104

Ogden, A. C. (2006). Ethnographic inquiry: Reframing the learning core of education abroad. *Frontiers: The Interdisciplinary Journal of Study Abroad, 8*, 87–112. https://doi.org/10.36366/frontiers.v13i1.175

Oxford, S. M. (2005). *Study abroad: Travel vacation in college.* Where in the World Publishing.

Palmer, N. W. (2015). Inverting the object of study: Recalibrating the frame of reference in study abroad experiences. *Teaching Theology and Religion, 18*(1), 63–72. https://doi.org/10.1111/teth.12264

Pennycook, A. (1994). *The cultural politics of English as an international language.* Longman.

Peterson, C. (2002). Preparing engaged citizens: Three models of experiential education for social justice. *Frontiers: The Interdisciplinary Journal of Study Abroad, 8*, 41–82.

Phillipson, R. (1992). *Linguistic imperialism.* Oxford University Press.

Plater, W. M., Jones, S. G., Bringle, R. G., & Clayton, P. H. (2009). Educating globally competent citizens through international service learning. In R. Lewin (Ed.), *The handbook of practice and research in study abroad: Higher education and the quest for global citizenship* (pp. 485–505). Routledge.

Raymond, E. M., & Hall, C. M. (2008). The development of cross-cultural (mis)understanding through volunteer tourism. *Journal of Sustainable Tourism, 16*(5), 530–514. https://doi.org/10.1080/09669580802159610

Rhoads, R. A., & Neururer, J. (1998). Alternative spring break: Learning through community service. *NASPA Journal, 35*(2), 100–118.
Rizvi, F. (2000). International education and the production of global imagination. In N. C. Burbules & C. A. Torres (Eds.), *Globalization and education: Critical perspectives* (pp. 205–226). Routledge.
Roberts, C., Byram, M., Barro, A., Jordan, S., & Street, B. (2001). *Language learners as ethnographers*. Multilingual Matters.
Roediger, D. (1991). *The wages of whiteness: Race and the making of the American working class*. Verso.
Saltmarsh, J. (1996). Education for critical citizenship: John Dewey's contribution to the pedagogy of community service learning. *Michigan Journal of Community Service Learning, 3*(1), 13–21. http://hdl.handle.net/2027/spo.3239521.0003.102
Skelly, J. M. (2009). Fostering engagement: The role of international education in the development of global civil society. In R. Lewin (Ed.), *The handbook of practice and research in study abroad: Higher education and the quest for global citizenship* (pp. 21–32). Routledge.
Skutnabb-Kangas, T., & Phillipson, R. (1989). "Mother tongue": The theoretical and sociopolitical construction of a concept. In U. Ammon (Ed.), *Status and function of languages and language varieties* (pp. 450–477). Walter de Gruyter. https://doi.org/10.1515/9783110860252.450
Streitwieser, B. T. (2009). Undergraduate research during study abroad: Scope, meaning, and potential. In R. Lewin (Ed.), *The handbook of practice and research in study abroad: Higher education and the quest for global citizenship* (pp. 399–419). Routledge.
Taylor, J. (2002). Metaphors we serve by: Investigating the conceptual metaphors framing national and community service and service-learning. *Michigan Journal of Community Service Learning, 9*(1), 45–57. http://hdl.handle.net/2027/spo.3239521.0009.105
Tiessen, R., & Huish, R. (2013). International experiential learning and global citizenship. In R. Tiessen & R. Huish (Eds.), *Globetrotting or global citizenship?* (pp. 3–20). University of Toronto Press.
Train, R. (2009). Toward a "natural" history of the native (standard) speaker. In N. M. Doerr (Ed.), *The native speaker concept: Ethnographic investigations of "native speaker effects"* (pp. 47–80). Mouton de Gruyter.
Tsing, A. (2000). The global situation. *Cultural Anthropology, 15*(3), 327–360. https://www.jstor.org/stable/656606

Urciuoli, B. (2016). The compromised pragmatics of diversity. *Language & Communication, 51*, 30–39. https://doi.org/10.1016/j.langcom.2016.07.005

Urciuoli, B. (2018). Introduction: Neoliberalizing undergraduate experience. In B. Urciuoli (Ed.), *The experience of neoliberal education* (pp. 1–16). Berghahn Books.

Vande Berg, M. (2009). Intervening in student learning abroad: A research-based inquiry. *Intercultural Education, 20*(1–2), 15–27. https://doi.org/10.1080/14675980903370821

Vande Berg, M., Connor-Linton, J., & Paige, R. M. (2009). The Georgetown consortium project: Interventions for student learning abroad. *Frontiers: The Interdisciplinary Journal of Study Abroad, 18*. https://doi.org/10.36366/frontiers.v18i1.251

Vološinov, V. N. (1973). *Marxism and the philosophy of language*. Harvard University Press.

Williamson, W. (2004). *Study abroad 101*. Agapy Publishing.

Woolf, M. (2007). Impossible things before breakfast: Myths in education abroad. *Journal of Studies in International Education, 11*(3–4), 496–509. https://doi.org/10.1177/1028315307304186

Woolf, M. (2010). Another mishegas: Global citizenship. *Frontiers: The Interdisciplinary Journal of Study Abroad, 19*, 47–60. https://doi.org/10.36366/frontiers.v19i1.273

Zemach-Bersin, T. (2009). Selling the world: Study abroad marketing and the privatization of global citizenship. In R. Lewin (Ed.), *The handbook of practice and research in study abroad: Higher education and the quest for global citizenship* (pp. 303–320). Routledge.

Index[1]

[1] Note: Page numbers followed by 'n' refer to notes.

L. S. Schedel, C. Jakubiak (eds.), *Voluntourism and Language Learning/Teaching*, Palgrave Advances in Language and Linguistics,
https://doi.org/10.1007/978-3-031-40813-7

The manufacturer's authorised representative in the EU is Springer
Nature Customer Service Centre GmbH, Europaplatz 3, 69115 Heidelberg,
Germany. If you have any concerns regarding our products, please
contact ProductSafety@springernature.com

Printed and bound by CPI Group (UK) Ltd, Croydon, CR0 4YY
01/10/2026
02225894-0012